2013—2025年国家辞书编纂出版规划项目
英汉信息技术系列辞书

**总主编 白英彩**

# A CONCISE ENGLISH-CHINESE COMPUTER NETWORK DICTIONARY

# 英汉计算机网络简明词典

主　编　章仁龙
主　审　张世永
副主编　（以姓氏笔画为序）
　　　　王小清　王　岱　王　文
　　　　石燕华　刘中原　陈志毅
　　　　洪　珂　钱向阳　曹玉声
　　　　董静翔

上海交通大学出版社

**内容提要**

　　本词典收录了计算机网络技术及其产业的研究、开发、应用和管理等方面的新词 10 000 条。词条均按英文字母顺序排列,并进行了规范和审定。本词典是目前国内收集计算机网络技术词汇最丰富、涉及面最广泛、内容最新的词典。本词典可为计算机网络技术及其相关专业的研究、开发应用和科技书刊编辑及文献译摘人员使用,也适合于非计算机网络技术专业人员及广大业余爱好者作为工具书学习参考之用。

**图书在版编目(CIP)数据**

英汉计算机网络简明词典/章仁龙主编. —上海:上海交通大学出版社,2017
ISBN 978-7-313-18155-8

Ⅰ.①英… Ⅱ.①章… Ⅲ.①计算机网络-词典-英、汉 Ⅳ.①TP393-61

中国版本图书馆 CIP 数据核字(2017)第 229796 号

**英汉计算机网络简明词典**

| | |
|---|---|
| 主　　编:章仁龙 | |
| 出版发行:上海交通大学出版社 | 地　　址:上海市番禺路 951 号 |
| 邮政编码:200030 | 电　　话:021-64071208 |
| 出 版 人:谈　毅 | |
| 印　　制:苏州市越洋印刷有限公司 | 经　　销:全国新华书店 |
| 开　　本:880mm×1230mm　1/32 | 印　　张:9.875 |
| 字　　数:376 千字 | |
| 版　　次:2017 年 11 月第 1 版 | 印　　次:2017 年 11 月第 1 次印刷 |
| 书　　号:ISBN 978-7-313-18155-8/TP | |
| 定　　价:98.00 元 | |

版权所有　侵权必究
告读者:如发现本书有印装质量问题请与印刷厂质量科联系
联系电话:0512-68180638

# 英汉信息技术系列辞书顾问委员会

**名誉主任** 何友声
　　　　　翁史烈
**主　任** 张　杰
**委　员**（以姓氏笔画为序）
　　　　　叶甜春　过敏意　朱三元　陈亚珠　邵志清
　　　　　何积丰　何新贵　沈昌祥　张　鳌　欧阳钟灿
　　　　　周兴铭　施伯乐　倪光南　梅　宏　韩英铎
　　　　　魏少军

# 英汉信息技术系列辞书编纂委员会

**名誉主任** 吴启迪

**名誉副主任** 赵建军

**主　　任** 胡铉亮　　**执行主任** 杨根兴

**副 主 任** 王珏明　黄国兴

**秘 书 长** 黄国兴（兼）

**副秘书长** 齐开悦　汪　镭　胡松凌

**委　　员**（以姓氏笔画为序）

马范援　王珏明　王思伟　王晓峰　王　杰　王　能
王豪行　白英彩　齐开悦　李光亚　李家滨　李明禄
杨根兴　陈卫东　沈忠明　邱卫东　张天蔚　张世永
张　轮　张冠男　谷大武　汪　镭　严晓华　林正浩
林福宗　姜岩峰　胡铉亮　胡松凌　俞　涛　顾君忠
敖青云　崔良沂　章仁龙　章玉宇　黄国兴　蒋思杰
蔡国钧　管海兵　藏燕阳　燕存正　薛　质　蔡立志

**系列辞书总主编**
　　白英彩

**副 总 主 编**
　　章仁龙　李家滨　王豪行

# 《英汉计算机网络简明词典》

# 编委会

**主　　任**　章仁龙

**副主任**　王　岱　董静翔　黄莎琳

**秘书长**　刘中原

**委　　员**（以姓氏笔画为序）

王小清　王　岱　王　文　石燕华
刘中原　陈志毅　钱向阳　章仁龙
曹玉声　黄莎琳　董静翔

# 序

信息技术(IT)这个词如今已广为人们知晓,它通常涵盖计算机技术、通信(含移动通信)技术、广播电视技术、以集成电路(IC)为核心的微电子技术和自动化领域中的人工智能(AI)、神经网络、模糊控制和智能机器人,以及信息论和信息安全等技术。

近 20 多年来,信息技术及其产业的发展十分迅猛。20 世纪 90 年代初,由信息高速公路掀起的 IT 浪潮以来,信息技术及其产业的发展一浪高过一浪,因特网(互联网)得到了广泛的应用。如今,移动互联网的发展势头已经超过前者。这期间还涌现出了电子商务、商务智能(BI)、对等网络(P2P)、无线传感网(WSN)、社交网络、网格计算、云计算、物联网和语义网等新技术。与此同时,开源软件、开放数据、普适计算、数字地球和智慧地球等新概念又一个接踵一个而至,令人应接不暇。正是由于信息技术如此高速的发展,我们的社会开始迈入"新信息时代",迎接"大数据"的曙光和严峻挑战。

如今信息技术,特别是"互联网+"已经渗透到国民经济的各个领域,也贯穿到我们日常生活之中,可以说信息技术无处不在。不管是发达国家还是发展中国家,人们之间都要互相交流,互相促进,缩小数字鸿沟。

上述情形映射到信息技术领域是:每年都涌现出数千个新名词、术语,且多源于英语。编纂委认为对这些新的英文名词、术语及时地给出恰当的译名并加以确切、精准的理解和诠释是很有意义的。这项工作关系到 IT 界的国际交流和大陆与港、澳、台之间的沟通。这种交流不限于学术界,更广泛地涉及 IT 产业界及其相关的商贸活动。更重要的是,这项工作还是 IT 技术及其产业标准化的基础。

编纂委正是基于这种认识,特组织众多专家、学者编写《英汉信息技术大辞典》《英汉计算机网络辞典》《英汉计算机通信辞典》《英汉信息安全技术辞典》《英汉三网融合技术辞典》《英汉人工智能辞典》《英汉建筑智能化技术辞典》《英汉智能机器人技术辞典》《英汉智能交通技术辞典》《英汉云计算·物联网·大数据辞

## 英汉计算机网络简明词典

典》《英汉多媒体技术辞典》和《英汉微电子技术辞典》，以及与这些《辞典》（每个词汇均带有释文）相对应的《简明词典》（每个词汇仅有中译名而不带有释文）共 24 册，陆续付梓。我们希望这些书的出版对促进 IT 的发展有所裨益。

这里应当说明的是编写这套书籍的队伍从 2004 年着手，历时 10 年，与时俱进的辛勤耕耘，终得硕果。他们早在 20 世纪 80 年代中期就关注这方面的工作并先后出版了《英汉计算机技术大辞典》（获得中国第十一届图书奖）及其类似的书籍，参编人数一直持续逾百人。虽然参编人数众多，又有些经验积累，但面对 IT 技术及其产业化如此高速发展，相应出现的新名词、术语之多，尤令人感到来不及收集、斟酌、理解和编纂之虞。如今推出的这套辞书不免有疏漏和欠妥之处，请读者不吝指正。

这里，编纂委尤其要对众多老专家执着与辛勤耕耘表示由衷的敬意，没有他们对事业的热爱，没有他们默默奉献的精神，没有他们追求卓越的努力，是不可能成就这一丰硕成果的。

在"英汉信息技术系列辞书"编辑、印刷、发行各个环节都得到上海交通大学出版社大力支持。尤其值得我们欣慰的是由上海交通大学和编纂委共同聘请的 12 位院士和多位专家所组成的顾问委员会对这项工作自始至终给予高度关注、亲切鼓励和具体指导，在此也向各位资深专家表示诚挚谢意！

编纂委真诚希望对这项工作有兴趣的专业人士给予支持、帮助并欢迎加盟，共同推动该工程早日竣工，更臻完善。

英汉信息技术系列辞书编纂委员会

名誉主任 吴启迪

2015 年 5 月 18 日

# 前　言

　　计算机网络是通信技术与计算机技术相结合的产物。自20世纪90年代中期以来,随着互联网的普及和广泛应用,互联网正以改变一切的力量,在全球范围掀起一场影响人类所有层面的深刻变革,"互联网十"正以雷霆万钧之势推进大众创业万众创新。网络技术日新月异的迅猛发展,时时涌现新技术专业术语、缩略语和新名词,特别是同一新技术名称和术语的中文译名和释义又往往会有多个不同的版本,有的差异还很大,造成理解和交流的困难。鉴于此,我们在编撰出版了《英汉信息技术大辞典》的基础上,编撰《英汉计算机网络辞典》这一专业分册。

　　本书编入了网络技术规范化名词术语10 000余条,内容涵盖网络架构、网络协议、网络管理、网络应用、网络安全和云计算等领域。书中所列名词术语及其缩略词译名,除以"全国科学技术委员会名词审定委员会"发布的为主要依据外,取约定俗成的或使用频率最高的作译名,在释义方面力求精确简明、浅显易懂,并努力反映其最新技术发展。以裨读者于网络技术的学习研究与应用。

　　在历时多年的编撰过程中,我们参阅了数以百计的相关文献、工具书、有关书籍和网站上刊载的信息,这里一并向各位作者表示诚挚的感谢!还要感谢英汉信息技术系列辞书总主编白英彩教授的悉心关心和具体指导。由于网络技术和产业发展迅速,加之学识有限,书中存在的疏漏之处,恳请广大读者不吝指正。

　　感谢深圳市普联技术公司董事长赵建军先生和网宿科技股份有限公司对本书出版的鼎力资助。

编　者

2017年8月

# 凡　例

1. 本词典按英文字母顺序排列,不考虑字母大小写。数字及希腊字母另排。专用符号(空格、圆点、连字符等)不参与排序。
2. 词汇的英文用粗体。一个英语词汇有多个译名时,可根据其彼此意义的远近用分号或逗号隔开。如"macro 宏命令,宏指令;宏"。
3. 圆括号( )内的内容表示解释或可以略去,如"system partition 系统(磁盘)分区"。
4. 方括号[ ]内的内容表示可以替换紧挨方括号前面的字词。如"Keyword 关键字[词]"。
5. 双页码上的书眉为本页第一个词汇的第一个单词;单页码上的书眉为本页最后一个词汇的第一个单词。
6. 对于英文名词术语的译名以全国科学技术委员会名词审定委员会发布的为主要依据,对于已经习惯的名词也作了适当反映,如"disk"采用"光碟"为第一译名,"光盘"为第二译名等。
7. 本词典中出现的计量大部分采用我国法定计量单位,但考虑读者查阅英文技术资料的方便,保留了少量英制单位。

# 网宿科技股份有限公司介绍

  网宿科技股份有限公司（股票代码：300017）成立于2000年，主要提供内容分发与加速（CDN）、互联网数据中心（IDC）、面向运营商的网络优化解决方案以及云计算服务，是国内专业的互联网业务平台提供商。公司紧跟互联网行业热点及发展趋势，不断研发创新，在服务集群优化、海量存储技术、内容路由技术、负载均衡技术、协议支持和优化、移动互联网、云计算、IPv6等领域进行了丰富的技术积累。通过自主研发，网宿不断推进技术创新、丰富产品线以及时满足市场需求，近年来逐步推出了针对不同行业的互联网加速以及云计算解决方案，通过覆盖全球的加速节点及海量的带宽储备，为国内外的众多著名客户提供了高品质的服务，大力促进了中国互联网产业的发展。

# 目 录

| | |
|---|---|
| A | 1 |
| B | 24 |
| C | 41 |
| D | 71 |
| E | 87 |
| F | 95 |
| G | 105 |
| H | 111 |
| I | 119 |
| J | 137 |
| K | 139 |
| L | 140 |
| M | 149 |
| N | 167 |
| O | 200 |
| P | 210 |
| Q | 229 |
| R | 231 |
| S | 240 |
| T | 261 |
| U | 272 |
| V | 277 |
| W | 284 |

X ………………………………………………………………… 294
Y ………………………………………………………………… 295
Z ………………………………………………………………… 296
以数字字母起首的词条 ………………………………………… 297

# A

AAA (Authentication, Authorization, Accounting) 认证,授权,计费
AAA server  AAA 服务器
AAFID (autonomous agent for intrusion detection) 自治代理入侵检测系统
AAL (ATM adaptation layer)  ATM 适配层,异步传输模式适配层
AAL connection  ATM 适配层连接
AAL service  ATM 适配层业务
AAL‑1 (ATM adaptation layer type 1)  ATM1 类适配层
AAL‑2 (ATM adaptation layer type 2)  ATM2 类适配层
AAL‑3/4 (ATM adaptation layer type 3/4)  ATM3/4 类适配层
AAL‑5 (ATM adaptation layer type 5)  ATM5 类适配层
. aam  Authorware 网络播放映射文件名后缀
AAMOF (as a matter of fact) 事实上
AARP (AppleTalk address resolution protocol)  AppleTalk 地址解析协议
AARP probe packets  AARP 探测报文分组
. aas  Authorware 网络传输片段文件名后缀
AASB (adaptive acceleration saves bandwidth) 自适应加速节省带宽

A&B bit signaling  A&B 位信令
abbreviated address calling 缩址呼叫
abbreviation name 简名
ABC (agent business consumer)  ABC 模式,代理商/商家/消费者
ABM (asynchronous balanced mode) 异步平衡模式
abnormal flow cleaning 异常流量清洗
abnormal detection 异常侦测
abnormal propagation 异常传播
abnormal propagation characteristics 异常传播特征
abnormal termination 异常结束
abnormal transmission 异常传输
A-Bone 亚太主干网
abort remirror 中止重镜像
above the line (ATL) 线上
ABR (available bit rate) 可用位速[比特]率
ABR (area border router) 区域边界路由器
ABRD (automatic baud rate detection) 自动波特率检测
abrupt release 鲁莽拆除
absolute phase shift keying (APSK) 绝对相移键控
absolute URL 绝对统一资源定位符
abstract data link layer 认证服务器抽

象数据链路层
abstract data model 抽象数据模型
abstract data model layer 抽象数据模型层
abstract data type (ADT) 抽象数据类型
abstract data type compiler 抽象数据类型编译器
abstract syntax notation (ASN) 抽象语法标记
abstract syntax tree (AST) 抽象语法树
ACAP (application configuration access protocol) 应用配置访问协议
acceptable level of risk 可接受的风险级别
acceptable use policy (AUP) 可接受使用政策
access amount counter 访问量计数器
access and control protocol 接入与控制协议
access attempt 接入试呼,访问尝试
access channel 接入信道
access channel control 接入[访问]信道控制(器)
access constraint 存取约束
access contention 接入争用
access contention resolution 接入争用裁决
access control 接入控制
access control byte 接入[访问]控制字节
access control channel 接入控制信道
access control entry (ACE) 接入[访问]控制项

access control mechanism 接入控制机制
access control strategy 接入控制策略
access counter 接入计数器
access delay 接入延迟
access delay time 接入时延,接入拒绝时间
access denial 接入拒绝
access denial probability 接入拒绝概率
access denial ratio 拒绝接入率
access denial time 接入拒绝时间
access function 接入函数
access granted channel (AGCH) 接入允许信道
access group 接入组群
access link control application part (ALCAP) 接入链路控制应用部分
access method 接入[访问]方法
access method control block 存取方法控制块
access network (AN) 接入网
access network authentication, authorization, accounting (AN-AAA) 接入网认证,授权,计费
access network control center 接入网控制中心
access network system management function (AN-SMF) 接入网系统管理功能
access node 接入节点
access path independence 接入路径独立性
access period 接入期限
access point (AP) 存取点,接入点

access priority　存取优先级
access priority control　存取优先控制
access privilege　存取权限
access procedure　存取规程
access protocol　接入协议
access provider　接入提供商
access rate (AR)　接入速率
access remote node　接入远程节点
access request (AR)　接入[访问]请求
access response channel　接入响应信道
access restriction enforcement　存取强制约束
access right　接入[访问]权限
access right constraint　存取权限约束
access server　访问服务器
access site　访问站点
access speed　访问速度
access stack node (ASN)　访问堆叠式节点
access termination　接入终端
access termination identifier (ATI)　接入终端标识
access termination unit remote (ATU-R)　远端接入端接单元
access time　接入时间
access time minimization　访问时间最小化
access token　访问令牌
access transportation network　接入传输网络
access unit (AU)　访问单元
access unit interface (AUI)　访问单元接口
access waiting time　接入等待时间
account　账户

accounting management (AM)　账目管理
ACDI (asynchronous communication device interface)　异步通信设备接口
ACE (access control entry)　接入[访问]控制项
ACGN (Animation Comic Game Novel)　二次元
ACI (adjacent channel interference)　相邻信道干扰
ACIA (asynchronous communication interface adapter)　异步通信接口适配器
ACK (acknowledge)　确认
ACK flood attack　ACK 泛洪攻击
ACKN (acknowledgment number)　确认号
acknowledged mail　应答邮件
acknowledged service　应答的服务
acknowledgement (ACK)　(应答)确认
acknowledgement frames　应答帧
acknowledgement number (ACKN)　确认号
acknowledgement signal　应答信号,确认信号
acknowledgement window　确认窗口
acknowledge number　应答号
ACK storm attack　ACK 风暴攻击
ACL (access control list)　访问控制列表
ACM (address complete message)　地址完成消息
ACO (colony optimization)　蚁群优化
ACPR (adjacent channel power ratio)　邻信道功率比

ACR (allowed cell rate) 允许的信元率
ACR decrease time factor 允许的信元率缩时因素
across certificate 交叉证书
ACS (advanced connectivity system) 先进布线系统
ACS (adjacent channel selectivity) 邻信道选择性
ACSE listing (association control service element) 关联控制服务元素
active attack 主动攻击
active backup link 活动备用链路
active black hole attack 主动型黑洞攻击
active channel 现用频道,工作信道
active channel state 工作信道状态
active congestion control 主动拥塞控制
active corrective network 有源校正网络
active crossover 有源分频
active crossover network 有源分频网络
active directory server 活动目录服务器
active directory service 活动目录服务
active directory service interface (ADSI) 活动目录服务接口
active directory tree 活动目录树
active distributed network 主动分布式网络
active distribution network 有源分配网
active document server 活动文档服务器
active double star (ADS) 有源双星
active equivalent network 有源等效网络
active event queue 活动事件队列
active filter network (AFN) 有源滤波器网络
active gateway 活跃网关
active hub 有源集线器
active linear network 有源线性网络
active link 活动链
active microwave network 有源微波网络
active microwave remote sensing 有源微波遥感
active microwave sensor 有源微波传感器
active network 有源网络
active network architecture 有源网络体系结构
active network congestion control (ANCC) 主动网络拥塞控制
active networking 主动网络
active node 主动节点
active open 主动打开
active optical device (AOD) 有源光器件
active page queue 活动页面队列
active parameter node 活动参数节点
active partitioning architecture 有源分隔体系结构
active queue management (AQM) 主动队列管理
active queue management algorithms 主动队列管理算法
active queue management scheme 主动

队列管理策略
active optical device　有源光器件
active optical fiber (AOF)　有源光纤
active optical network (AON)　有源光网络
active repeater　有源中继器
active routing algorithm　主动路由选择算法
active routing protocol　主动路由协议
active server　活动服务器
active server pages (ASP)　活动服务器页面
active signalling link　活动信号链路
active star　有源星形
active star coupler　有源星形耦合器
active star network　有源星形网
active star topology　有源星形拓扑
active station　主动[活动]站
active streaming format (ASF)　活动流格式
active structure network (ASN)　活动结构网络
active virtual forwarder (AVF)　活跃虚拟转发器
active virtual gateway (AVG)　活跃虚拟网关
active virtual server　主动虚拟服务器
active virtual terminal　活跃虚拟终端
active Web GIS　主动的 Web 地理信息系统
active wire tapping　主动搭线窃听
activity list　活动表
activity management　活动管理
activity node network　活动结点网络
actual maximum used routing buffers 实际最多使用的路由缓存数
AD (administrative distance)　管辖距离
AD (automatic deployment)　自动部署
adaptive acceleration　自适应加速
adaptive acceleration model　自适应加速模型
adaptive acceleration saves bandwidth (AASB)　自适应加速节省带宽
adaptive algorithm　自适应算法
adaptive backoff algorithm　自适应退让算法
adaptive blind equalization　自适应盲均衡
adaptive bridge　自适应网桥
adaptive channel　自适应信道
adaptive channel allocation　自适应信道分配
adaptive channel blind equalization　自适应信道盲均衡
Adaptive channel coding　自适应信道编码
adaptive channel equalization　自适应信道均衡
adaptive complex channel equalization　自适应复合信道均衡
adaptive cut-through switching　自适应直通交换
adaptive digital equalization　自适应数字均衡
adaptive directory　自适应目录
adaptive distributed clustering routing　自适应分布式聚簇路由协议
adaptive distributed computing　自适应分布计算

| adaptive distributed control 自适应分布式控制
| adaptive dynamic algorithm 自适应动态学习算法
| adaptive dynamic analysis 自适应动态分析
| adaptive dynamic arbiter 自适应动态仲裁器
| adaptive dynamic fuzzy system 自适应动态模糊系统
| adaptive dynamic matching 自适应动态匹配
| adaptive dynamic programming 自适应动态规划
| adaptive dynamic protecting 自适应动态保护
| adaptive dynamic routing algorithm 自适应动态路由算法
| adaptive dynamic threshold 自适应动态阈值
| adaptive equalization algorithm 自适应均衡算法
| adaptive equalizer (AE) 自适应均衡器
| adaptive channel estimator 自适应信道估计
| adaptive fault detection 自适应故障检测
| adaptive fault feeder selection 自适应故障选线
| adaptive fault tolerance (AFT) 自适应性容错
| adaptive fault tolerant routing 自适应容错路由
| adaptive fault topology structure 自适应故障拓扑结构

adaptive filtering algorithm 自适应滤波算法
adaptive frequency hopping (AFH) 自适应跳频
adaptive fuzzy fault tolerant control 自适应模糊容错控制
adaptive fuzzy net 自适应模糊网
adaptive immune response network (AIRN) 自适应免疫应答网络
adaptive joint equalization 自适应联合均衡
adaptive load balancing (ALB) 自适应负载均衡
adaptive multipath routing protocol 自适应多径路由协议
adaptive opportunistic channel access (AOCA) 自适应机会信道接入
adaptive path index (APEX) 自适应的路径索引
adaptive reconstruction algorithm 自适应重建算法
adaptive retransmission 适应性重传
adaptive retransmission algorithm 自适应重传算法
adaptive robust fault tolerant control 自适应鲁棒[健壮]容错控制
adaptive routing (AR) 自适应路由选择
adaptive session-level pacing 适应性会话级[层]定步
adaptive time algorithm 自适应时间算法
adaptive tree walk protocol 自适应树步进协议
adaptive watermarking algorithm 自适

应水印算法
ADCCP（advanced data communication control procedures） 高级数据通信控制规程
additional header information 附加报头信息
additive increase rate（AIR） 加法增加率
address complete message（ACM） 地址完成消息
address depletion 地址消耗
address family border router 地址簇边界路由器
address field 地址字段
address field extension 地址字段扩充
address head 地址标题
address indicating group（AIG） 地址指示组
address indicating group allocation 地址指示组分配
address mapper 地址映像器
address mapping 地址映射[映像]
address mapping external sorting 地址映射外分类
address mask 地址掩码
address mask reply 地址掩码响应
address mask request 地址掩码请求
address message 地址消息
address message sequencing 地址报文排序
address prefix 地址前缀
address recognized indicator/frame copied indicator（ARI/FCI） 地址已识别指示符/帧复制指示符
address resolution 地址解析

address resolution cache 地址转换高速缓存
address resolution protocol（ARP） 地址解析协议
address resolution protocol request（ARP request） 地址解析协议请求
address sequencing 地址排序法
address space 地址空间
address space manager（ASM） 地址空间管理器
address space mapping 地址空间映射
address spoofing 地址欺诈
address table mapping 地址表映射
address template protection 地址模板保护
address translation gateway（ATG） 地址转换网关
add water 灌水
Ad hoc network 自组织网络，暂时专用网络
Ad hoc on-demand distance vector routing（AODV） 按需距离向量路由选择
Ad hoc protocol 自组织网络协议
Ad hoc WLAN 自组织无线局域网
adjacent and alternate-channel selectivity 相邻和间隔的信道选择性
adjacent channel 相邻信道
adjacent channel attenuation 相领信道衰减
adjacent channel interference（ACI） 相邻信道干扰
adjacent channel power ratio（ACPR） 邻信道功率比
adjacent channel selectivity（ACS） 邻

信道选择性
adjacent control point 邻近控制点
adjacent domains 相邻区域
adjacent frequency channel transfer 间隔信道传输
adjacent link station 邻接链路站
adjacent link station image 邻接链路站映像
adjacent link storage image 相邻链路存储映像
adjacent network 伴随网络
adjacent node 相邻节点
adjacent node algorithm 相邻节点算法
adjacent subarea 相邻子区
adjacent subnetwork 相邻子网
adjunct service point (ASP) 附加业务点
adjunct network 附属网络
ADM (asynchronous disconnected mode) 异步断路模式
administrative alerts 管理报警信号
administrative distance (AD) 管辖距离
administrative domain 管理域
administrator (ADMIN) 系统管理员
admission delay 允许延迟
ADSI (active directory service interface) 活动目录服务接口
ADSL (asynchronous digital subscriber loop) 非对称数字用户环路
ADSL (asymmetric digital subscriber line) 非对称数字用户线路
ADSL transceiver unit (ATU) ADSL 收发单元

ADSL transceiver unit, central office end (ATU-C) ADSL 局端收发单元
ADSL transceiver unit, remote terminal end (ATU-T) ADSL 远端收发单元
ADSP (AppleTalk data stream protocol) AppleTalk 数据流协议
ADSU (ATM data service unit) ATM 数据服务单元
ADT (asynchronous data transmission) 异步数据传输
ADT (abstract data type) 抽象数据类型
advanced communication riser (ACR) 高级通信插卡
advanced communication service (ACS) 高级通信业务(网)
advanced communication system (ACS) 高级通信系统
advanced connectivity system (ACS) 先进布线系统
advanced data communication control procedure (ADCCP) 高级数据通信控制规程
advanced data connector (ADC) 高级数据连接器
advanced data link control (ADLC) 高级数据链路控制(协议)
advanced encryption standards (AES) 高级加密标准
advanced intelligent network (AIN) 高级智能网络
advanced net 先进网络
advanced network system architecture (ANSA) 先进网络系统结构
advanced peer-to-peer communication

（APPC） 高级对等通信
advanced peer-to-peer networking (APPN) 高级对等联网，高级点对点网络
advanced peer-to-peer networking (APPN) end node 高级对等联网终结节点
advanced peer-to-peer networking (APPN) interchange node 高级对等联网交换节点
advanced peer-to-peer networking (APPN) low entry node 高级对等联网低级入口节点
advanced peer-to-peer networking (APPN) network node 高级对等联网网络节点
advanced program-to-program communication (APPC) 高级程序间通信
Advanced Research Projects Agency (ARPA) 高级研究计划局
Advanced Research Project Agency network (ARPAnet) 阿帕网，高级研究计划局网
advanced streaming format (ASF) 高级流格式
advertising 广告，发布
advertising mail 广告邮件
AdvTHANKSance (Thanks In Advance) 预先感谢
adware 广告软件
AE (application entity) 应用实体
AE (adaptive equalizer) 自适应均衡器
AEP (AppleTalk echo protocol) AppleTalk 回声协议
AES (advanced encryption standards) 高级加密标准
AFAIK (as far as I know) 据我所知
affiliate marketing 联盟营销
affinity-based routing 亲缘路由选择
affirmation acknowledgement (ACK) 确认收到
AFH (adaptive frequency hopping) 自适应跳频
AFK (away from keyboard) 离开键盘，暂时离开
AFN (active filter network) 有源滤波器网络
Africa Internet Network Information Center (AfriNIC) 非洲互联网络信息中心
AFT (adaptive fault tolerance) 自适应性容错
AG (application gateway) 应用网关
AGCH (access granted channel) 接入允许信道
agent business consumer (ABC) ABC 模式，代理商/商家/消费者
agent transfer protocol (ATP) 代理传输协议
aggregatable global unicast address 可聚合全球单播地址
aggregate bandwidth 聚合带宽
aggregate class website 聚合类网站
aggregate port (AP) 聚集端口
aggregate route-based IP switching (ARIS) 基于聚集路由 IP 交换方式
aggregator 聚合器
AGN (auto-detect global network) 全自动识别全球网络
agricultural virtual reality technology

农业虚拟现实技术
AGP (application gateway proxy) 应用网关代理
agreement on basic telecommunications 基础电信协议
AH (authentication header) 鉴别报头
AHFG (ATM-attached host functional group) ATM 连接的宿主功能组
AHS (application hosting service) 应用主机服务
air interface 空中接口
airmail 移动邮件
AIRN (adaptive immune response network) 自适应免疫应答网络
AirPort 空港方案
AirSnort AirSnort 工具,一种黑客工具
AKA (also known as) 也称
ALAP (AppleTalk link access protocol) AppleTalk 链路层访问协议
alarm indication signal (AIS) 报警指示信号
ALB (adaptive load balancing) 自适应负载均衡
ALCAP (access link control application part) 接入链路控制应用部分
ALG (application level gateway) 应用层网关
ALGS (application layer gateway service) 应用层网关服务
alias mail box 中转信箱
alias network address 别名网络地址
Alipay 支付宝
Alipay account 支付宝账户
all channel communication network 全通道通信网络
all IP universal mobile telecommunication system (IP-UMTS) 全 IP 通用移动通信系统
allocated circuit 分配式线路
all optical communication and sensing network 全光通信和传感网络
all optical communication network 全光通信网络
all optical communication system 全光通信系统
all optical fiber communication 全光纤通信
all optical fiber communication network 全光纤通信网络
all optical network (AON) 全光网络
all optical wave converter (AOWC) 全光波长转换器
allowable packet dropout rate 容许数据包丢失率
allowed cell rate (ACR) 允许的信元率
all-pass network 全通网络
all routes broadcast frame 全路由广播帧
all routes busy (ARB) 路由全忙
all routes explorer frame 全路由探测帧
all routes explorer packet 全路由探测数据包
all-station broadcast frame 全站广播帧
all-stations address 所有站地址
all trunks busy (ATB) 所有中继线忙
all 1's broadcast address 全 1 广播地址

ALP (application layer protocol) 应用层协议
also known as (AKA) 也称
alternate circuit-switched voice/circuit-switched data (CSV/CSD) 候选电路交换话音与电路交换数据
alternate communication net 备用通信网
alternate extended route 替代扩充路由
alternate mark inversion (AMI) 传号交替反转
alternate mark inversion (AMI) signal 传号交替反转信号
alternate route 替换路由
alternate route pattern 替换路由模式,替代路由图
alternative frequency 替换[备用]频率
alternative line 替换线路
alternative newsgroup hierarchy 可选新闻组等级结构
alternative routine 迂回路由选择
alternative routing indicator 选用路由选择指示符
always on 永远连网(状态)
AM (amplitude modulation) 幅值调制,调幅
AM (accounting management) 账目管理
AM (application monitoring) 应用监控
ambient ubiquity network (AUN) 环境感知泛在网络
American Online (AOL) 美国在线,美国联机服务

American Registry for Internet Numbers (ARIN) 美国因特网号注册机构
amoeba server 变形虫服务器
amplification attack 放大攻击
amplitude modulation (AM) 幅值调制,调幅
amplitude phase keying (APK) 幅相键控
amplitude pulse code modulation (APCM) 幅度脉码调制
amplitude quantize control 幅度量化控制
amplitude shift keying (ASK) 幅移键控
amplitude shift modulation 幅移调制
AN-AAA (access network authentication, authorization, accounting) 接入网认证,授权,计费
analog front end 模拟前端
analog integrated circuit neural network 模拟集成电路神经网络
analog network 模拟网络
analog simultaneous voice and data (ASVD) 模拟语音数据同传
analysis model of network behaviors 网络用户行为分析模型
ANCC (active network congestion control) 主动网络拥塞控制
anchor 锚点
anchored graphic 锚定图形
anchored layout 锚点式布局
anchored object 锚点对象
anchor text 锚文本
anchor text hiberarchy 锚文本层次结构

## A

anchor text links 锚文本链接
anchor text optimization 锚文本优化
Animation Comic Game Novel (ACGN) 二次元
anisochronous signal 准同步信号
anisochronous transmission 准同步传输,非等时同步传输
anomalous propagation (AP) 异常传播
anomalous propagation characteristics 异常传播特征
anonymous access 匿名访问
anonymous access and authentication control 匿名访问和身份验证控制
anonymous attack 匿名攻击
anonymous electronic mail 匿名电子邮件
anonymous file transfer protocol 匿名文件传送协议
anonymous logon 匿名注册
anonymous network 匿名网络
anonymous network communication 匿名网络通信
anonymous network connection 匿名网络连接
anonymous pipe 无名[匿名]管道
anonymous post 匿名发送
anonymous proxy 匿名代理
anonymous proxy server 匿名代理服务器
anonymous remailer 匿名邮件转送
anonymous routing 匿名路由
anonymous routing protocol 匿名路由协议
anonymous server 匿名服务器
anonymous speech 匿名言论

anonymous user access 匿名用户访问
anonymous users 匿名用户
anonymous Web browsing 匿名网页浏览
ANSA (advanced network system architecture) 先进网络系统结构
ANSA (active networking security architecture) 主动网络安全体系结构
answer message (ANM) 回答消息
answer mode 应答模式
answer seizure ratio (ASR) 应答占用率
ant colony optimization (ACO) 蚁群优化
ant colony system (ACS) algorithm 蚁群算法
anti-phishing 反网络钓鱼
Anti-Phishing Alliance of China (APAC) 中国反钓鱼网站联盟
Anti-Phishing Working Group (APWG) 反网络钓鱼工作组
anti-replay attack 反重放攻击
anti-replay mechanism 反重放机制
anti-replay protocol 反重放协议
anti-spam gateway 反垃圾邮件网关
anti-spam laws 反垃圾邮件法
anti-spam products 反垃圾邮件产品
anti-spyware coalition (ASC) 反间谍软件联盟
antisymmetric cryptology 非对称型密码
anycast 任播
anycast address 任播地址
anycast address resolution protocol (AARP)

任播地址解析协议
anycast flow　任波流
anycast policy　选播策略
anycast router　选播路由器
any source multicast (ASM)　任意源组播
any to any (A2A)　任何到任何
any to any connectivity　任意互连
any to any multicast　任意组播
AOCA (adaptive opportunistic channel access)　自适应机会信道接入
AOD (active optical device)　有源光器件
AODV (Ad hoc on-demand distance vector routing)　按需距离向量路由选择
AOF (active optical fiber)　有源光纤
AON (all optical network)　全光网络
AON (active optical network)　有源光网络
AOWC (all optical wave converter)　全光波长转换器
AP (access point)　存取点，接入点
AP (aggregate port)　聚集端口
APAC (Anti-Phishing Alliance of China)　中国反钓鱼网站联盟
Apache group　Apache 集团
Apache server　Apache 服务器
Apache Software Foundation (ASF)　Apache 软件基金会
APaRT (automated packet recognition/translation)　自动分组识别/转换
APCERT (Asian Pacific Computer Emergency Response Team)　亚太地区计算机应急响应组
APCM (amplitude pulse code modulation)　幅度脉码调制
APE (application protocol entity)　应用协议实体
API (application programming interface)　应用程序编程接口
APIPA (automatic private IP addressing)　专用 IP 自动寻址
APIX (Asian Pacific Internet Exchange)　亚太互联网交换中心
.apk (Android Package)　Android 安装包文件名后缀
APK (amplitude phase keying)　幅相键控
APNIC (Asian Pacific Network Information Center)　亚太网络信息中心
APON (ATM passive optical network)　异步传输模式无源光网络
App (application)　应用程序
APPC (advanced program-to-program communication)　高级程序间通信
APPC (advanced peer-to-peer communication)　高级对等通信
AppDrvN (application driven networking)　应用驱动联网
APPL (application program)　应用程序
Apple Pay　苹果手机移动支付,苹果支付
Applet　小应用程序
AppleTalk　AppleTalk 网,Apple 计算机网络协议
AppleTalk address resolution protocol (AARP)　AppleTalk 地址解析协议
AppleTalk datagram delivery protocol (DDP)　AppleTalk 数据报传送协议
AppleTalk data stream protocol (ADSP)　AppleTalk 数据流协议

## A

AppleTalk echo protocol (AEP) AppleTalk 回声协议
AppleTalk filing protocol (AFP) AppleTalk 文件协议
AppleTalk high layer protocol　AppleTalk 高层协议
AppleTalk link access protocol (ALAP) AppleTalk 链路层访问协议
AppleTalk network architecture　AppleTalk 网络体系结构
AppleTalk network layer　AppleTalk 网络层
AppleTalk session protocol (ASP) AppleTalk 会话协议
AppleTalk transaction protocol (ATP) AppleTalk 交易协议
AppleTalk transport layer protocols AppleTalk 传输层协议
AppleTalk update-based routing protocol (AURP)　AppleTalk 基于更新的路由选择协议
application accelerator (APPA)　应用加速
application compatibility layer　应用兼容层
application configuration access protocol (ACAP)　应用配置访问协议
application configuration files　应用(程序)配置文件
application configuration interface　应用配置接口
application context　应用上下文
application driven networking (AppDrvN)　应用驱动联网
application entity (AE)　应用实体

application entity address　应用实体地址
application entity type　应用实体类型
application exchange platform (AXP) 应用交换平台
application framework layer　应用框架层
application gateway (AG)　应用网关
application gateway proxy (AGP)　应用网关代理
application hosting service (AHS)　应用主机服务
application integration middleware　应用集成中间件
application intelligence service　应用智能服务
application interface layer　应用接口层
application layer　应用层
application layer gateway (ALG)　应用层网关
application layer gateway service (ALGS)　应用层网关服务
application layer interface　应用层接口
application layer middleware　应用层中间件
application layer multicast　应用层组播
application layer multicast protocol　应用层组播协议
application layer protocol (ALP)　应用层协议
application layer protocols analysis　应用层协议分析
application layer protocol filter　应用层协议过滤

application layer protocol specification 应用层协议规范
application layer routing 应用层路由
application layer service definition 应用层服务定义
application level 应用级
application level filtering 应用级过滤
application level framing (ALF) 应用级组帧
application level gateways 应用层网关
application level proxy 应用级代理
application middleware 应用中间件
application monitoring (AM) 应用监控
application network 应用网络
application programming interface (API) 应用程序编程接口
application protocol 应用协议
application protocol entity (APE) 应用协议实体
application scope 应用程序作用域
application server 应用服务器
application service element (ASE) 应用服务元素
application service provider (ASP) 应用服务提供商
application specific iIntergrated circuit (ASIC) 专用集成电路
application support layer 应用支撑层
application transparency 应用透明度
application vulnerability description language (AVDL) 应用脆弱性描述语言
App marketing App 营销
APPN (advanced peer-to-peer networking) 高级对等联网
APPN intermediate routing 高级对等联网中间路由选择
App Store mode 应用商店模式
APSK (absolute phase shift keying) 绝对相移键控
APWG (Anti-Phishing Working Group) 反网络钓鱼工作组
AQA (any question answered) 你问我答
AQM (active queue management) 主动队列管理
AR (augmented reality) 增强现实
ARB (all routes busy) 路由全忙
arbitrated loop 仲裁环
arbitrated signature 仲裁签名
Archie Archie 搜索引擎,阿奇
Archie client Archie 客户程序
Archie server Archie 服务器(程序)
Archie user interfaces Archie 用户接口
archive polling service 文档查询服务
archive server 文档服务器
area border router (ABR) 区域边界路由器
area resource management (ARM) 区域资源管理
ARIB (Association of Radio Industries and Businesses) (日本)无线工业及商贸联合会
ARIN (American Registry for Internet Numbers) 美国因特网号注册机构
ARM (asynchronous response mode) 异步响应模式
ARM (area resource management) 区域资源管理

ARP (address resolution protocol) 地址解析协议

ARPA (Advanced Research Projects Agency) （美国国防部的）高级研究计划局

ARPAnet (Advanced Research Projects Agency network) 阿帕网，（美国国防部的）高级研究计划局网

ARPA network elements and operation ARPA 网络元素和运行

ARPA network protocol level ARPA 网络协议级

ARP request (address resolution protocol request) 地址解析协议请求（报文）

ARP spoofing ARP 欺诈

ARQ (automatic repeat request) 自动重发请求

ARQ equipment 自动重发请求设备

arrangement unit (AU) 管理单元

artificial cognition calculability model 人工认知可计算模型

artificial immune network 人工免疫网络

artificial immune network memory 人工免疫网络记忆

artificial immune neural network 人工免疫神经网络

artificial intellectual network 人工智能网络

artificial intellectual neural network 人工智能神经网络

artificial network 仿真网络

artificial neural nets (ANN) 人工神经网络

artificial neural network algorithm 人工神经网络算法

artificial neural network control 人工神经网络控制

artificial neural network ensembles 人工神经网络集成

AS (authentication server) 认证服务器

AS (autonomous system) 自治系统

ASAP (as soon as possible) 尽快

ASBR (autonomous system boundary router) 自治系统边界路由器

ASC (anti-spyware coalition) 反间谍软件联盟

.asc ASCII 字符文件名后缀

ASCII (American Standard Code for Information Interchange) 美国信息交换标准代码

ASCII (American Standards Committee for Information Interchange) 美国信息交换标准委员会

ASCII art ASCII 艺术

ASCII protocol ASCII 协议

ASCII text ASCII 文本

ASCII value ASCII 值

ASDI (automated selective dissemination of information) 信息自动选择传输

ASE (application service element) 应用服务元素

ASF (Apache Software Foundation) Apache 软件基金会

ASF (advanced streaming format) 高级流格式

ASF (active streaming format) 活动流（式）格式

as far as I know (AFAIK) 据我所知

ASI (asynchronous serial interface) 异

步串行接口

Asian Pacific Computer Emergency Response Team (APCERT)  亚太地区计算机应急响应组

Asian Pacific Internet Exchange (APIX)  亚太互联网交换中心

Asian Pacific Network Information Center (APNIC)  亚太网络信息中心

ASIC (application specific iIntergrated circuit)  专用集成电路

ASK (amplitude shift keying)  幅移键控

askwitkey  问答型威客

ASM (address space manager)  地址空间管理器

ASM (any source multicast)  任意源组播

ASN (autonomous system number)  自治系统号

ASN (abstract syntax notation)  抽象语法标记

ASN (access stack node)  访问堆叠式节点

ASON (automatic switched optical network)  自动交换光网络

ASON-CP (automatic switched optical network control plane)  自动交换光网络控制平面

ASON-MP (automatic switched optical network management plane)  自动交换光网络管理平面

ASON-PC (automatic switched optical network permanent connection)  自动交换光网络永久连接

ASON-SC (automatic switched optical network switched connection)  自动交换光网络交换连接

ASON-SPC (automatic switched optical network soft permanent connection)  自动交换光网络软永久连接

ASON-TP (automatic switched optical network transport plane)  自动交换光网络传送平面

ASP (application service provider)  应用服务提供商

ASP (active server pages)  活动服务器页面

ASP (adjunct service point)  附加业务点

ASP (AppleTalk session protocol)  AppleTalk 会话协议

.asp  活动服务器页面文件名后缀

ASP.NET server control  ASP.NET 服务器控件

ASP.NET Web application  ASP.NET 万维网应用程序

ASR (answer seizure ratio)  应答占用率

asset of information service system  信息服务业务系统资产

asset of instant messaging system  即时消息业务系统资产

asset of the domain name registration system  域名注册系统资产

asset value of information service system  信息服务业务系统资产价值

asset value of instant messaging system  即时消息业务系统资产价值

asset value of the domain name registration

system 域名注册系统资产价值
assign a channel 分配通道
assigned access name 赋值的访问名
assigned cell 赋值的信元
association control channel (ACCH) 随路控制信道
association control protocol machine 联系控制协议机
association control service element (ACSE) 关联控制服务元素
Association of Radio Industries and Businesses (ARIB) （日本）无线工业及商贸联合会
AST (abstract syntax tree) 抽象语法树
A star algorithm A 星算法
asymmetrical channel 不对称信道
asymmetrical modulation 非对称调制
asymmetrical transmission 不对称传输，异步传输
asymmetric digital subscriber line (ADSL) 非对称数字用户线路
asymmetric digital subscriber line network 非对称数字用户线网络
asymmetric digital subscriber loop (ADSL) 非对称数字用户环路
asymmetric digital watermark 非对称数字水印
asymmetric network equilibrium problems (ANEP) 非对称网络均衡问题
asymmetric network 非对称网络
asymmetric network effect 非对称网络效应
asymmetric network information 非对称网络信息

asymmetric token bus and token ring network 非对称令牌环网络
asymmetric virtual local area network 非对称虚拟局域网
asynchronous attack 异步攻击
asynchronous balanced mode extended (ABME) 异步平衡模式扩展
asynchronous communication 异步通信
asynchronous communication device interface (ACDI) 异步通信设备接口
asynchronous communication interface adapter (ACIA) 异步通信接口适配器
asynchronous communication network 异步通信网
asynchronous data communication 异步数据通信
asynchronous data communication protocol 异步数据通信协议
asynchronous data transmission (ADT) 异步数据传输
asynchronous disconnected mode (ADM) 异步断路模式
asynchronous gateway 异步网关
asynchronous interactive Web environment 异步交互 Web 环境
asynchronous JavaScript And XML (AJAX) 异步 JavaScript 和 XML
asynchronous learning network 异步学习网络
asynchronous network 异步网络
asynchronous network model 异步网络模型

asynchronous neural network 异步类神经网络

asynchronous serial communication 异步串行通信

asynchronous series communication interface 异步串行通信接口

asynchronous sequential network 异步顺序网络

asynchronous serial interface (ASI) 异步串行接口

asynchronous telephone network 异步电话网

asynchronous time-division multiplexing (ATDM) 异步时分复用

asynchronous transfer mode (ATM) 异步传输模式

asynchronous transfer mode (ATM) network 异步传输模式网

asynchronous Web requests 异步Web请求

asynchronous Web services 异步Web服务

asynchronous Web services proxy 异步Web服务代理

asynchronous wireless mesh network 异步无线网状网

aTdHvAaNnKcSe (thanks in advance) 预先感谢

ATDM (asynchronous time-division multiplexing) 异步时分复用

ATG (address translation gateway) 地址转换网关

ATI (access termination identifier) 接入终端标识

ATL (above the line) 线上

ATM (asynchronous transfer mode) 异步传输模式

ATM access network ATM接入网

ATM adaptation layer (AAL) 异步传输模式适配层

ATM adaptation layer type 1 (AAL-1) ATM1类适配层

ATM adaptation layer type 2 (AAL-2) ATM2类适配层

ATM adaptation layer type 3/4 (AAL-3/4) ATM3/4类适配层

ATM adaptation layer type 5 (AAL-5) ATM5类适配层

ATM address ATM地址

ATM-attached host functional group (AHFG) ATM连接的宿主功能组

ATM backbone network ATM骨干网

ATM-based passive optical network 基于ATM的无源光网络

ATM cell ATM信元

ATM connection ATM连接

ATM data service unit (ADSU) ATM数据服务单元

ATM data network ATM数据网络

ATM exchange network ATM交换网

ATM forum ATM论坛

ATM LAN emulation ATM局域网仿真

ATM layer ATM(协议)层

ATM layer link ATM层链路

ATM link ATM链路

ATM management interface ATM管理接口

ATM management object ATM管理对象

**ATM management network** ATM 管理信息网

**ATM network integrated processing (ATM-NIP)** ATM 网络综合处理

**ATM network interface** ATM 网络接口

**ATM network interface card (ATM NIC)** ATM 网络接口卡

**ATM-NIC (ATM network interface card)** ATM 网络接口卡

**ATM-NIP (ATM network integrated processing)** ATM 网络综合处理

**ATM passive optical network (ATM-PON, APON)** 异步传输模式无源光网络,ATM 无源光网络

**ATM physical layer** ATM 物理层

**ATM-PON (ATM passive optical network)** 异步传输模式无源光网络

**ATM primitives** ATM 基本程序单元

**ATM protocol layers** ATM 协议层次

**ATM switch-to-switch interface (SSI)** ATM 交换机到交换机接口

**ATM user network interface (UNI)** ATM 用户网络接口

**ATM user-user connection** ATM 用户-用户连接

**ATP (AppleTalk transaction protocol)** AppleTalk 交易协议

**ATP (agent transfer protocol)** 代理传输协议

**attached resource computer network (ARCnet)** ARCnet 网,附加资源计算机网络

**attaching device** 连网设备

**attachment unit interface (AUI)** 连接单元接口

**attack** 攻击

**attack registry & intelligence service (ARIS)** 攻击事件注册及智能服务

**ATU (ADSL transceiver unit)** ADSL 收发单元

**ATU-C (ADSL transceiver unit, central office end)** ADSL 局端收发单元

**ATU-R (ADSL transceiver unit, remote terminal end)** ADSL 远端收发单元

**AU (arrangement unit)** 管理单元

**audio-video interactive service (AVIS)** 音频/视频交互服务

**audio-video interleaved (AVI)** 音频/视频交替格式

**augmented reality (AR)** 增强现实

**augmented reality gaming** 增强现实游戏

**augmented reality online gaming** 增强现实网络游戏

**augmented reality system** 增强现实系统

**augmented virtual reality** 增强虚拟现实

**augmented virtual reality technology** 增强虚拟现实技术

**AUI (attachment unit interface)** 连接单元接口

**AUN (ambient ubiquity network)** 环境感知泛在网络

**AUP (acceptable use policy)** 可接受使用政策

**AURP (AppleTalk update-based routing protocol)** AppleTalk 基于更新的路由选择协议

Authentication, Account, Authorization, Audit (AAAA, 4A) 认证,账号,授权,审计

Authentication, Authorization, Accounting (AAA, 3A) 认证,授权,计费

authentication exchange 验证交换

authentication header (AH) 鉴别报头

authentication of users 用户鉴别

authentication history server 认证历史服务器

authentication server (AS) 认证服务器

authenticator 身份验证者

authentic transmit mode 认证传输模式,可信传输模式

authorized APPN end node 授权的高级对等联网终端节点

authorized entry 授权登录

auto-detect global network (AGN) 全自动识别全球网络

AutoIP (automatic Internet protocol addressing) 自动IP寻址技术

auto-logon 自动注册

auto-logout 自动退出登录

automatically switched optical network (ASON) 自动交换光网络

automatically switched optical network control plane (ASON-CP) 自动交换光网络控制平面

automatically switched optical network management plane (ASON-MP) 自动交换光网络管理平面

automatically switched optical network permanent connection (ASON-PC) 自动交换光网络永久连接

automatically switched optical network soft permanent connection (ASON-SPC) 自动交换光网络软永久连接

automatically switched optical network switched connection (ASON-SC) 自动交换光网络交换连接

automatically switched optical network transport plane (ASON-TP) 自动交换光网络传送平面

automatically switched transport networks 自动交换传送网络

automatic alternate routing 自动转接[替换]路由选择

automatic baud rate detection (ABRD) 自动波特率检测

automatic call reconnect 自动请求重连

automatic catalog search 自动目录搜索

automatic collision detection 自动碰撞检测

automatic configuration link aggregation 自动配置链路聚合

automatic deployment (AD) 自动部署

automatic deployment of application 自动应用部署

automatic dialup 自动拨号连网

automatic forwarding 自动转发

automatic identification hyperlinks 自动识别超连接

automatic Internet protocol addressing (AutoIP) 自动IP寻址技术

automatic laser shutdown (ALS) 自动激光关闭

automatic link establishment (ALE) 自动链路建立

automatic link establishment protocol 自动链路建立协议
automatic link maintenance 自动链路保持
automatic optical-path switching protection 自动光路切换保护
automatic path generation 自动路径生成
automatic path management 自动路径管理
automatic path planning 自动路径规划
automatic poll 自动探询[轮询]
automatic private IP addressing (APIPA) 专用IP自动寻址
automatic purge/copy/redirect 自动清除、复制与改向
automatic request repeat 自动请求重复
automatic repeat request (ARQ) 自动重发请求
automatic repeat request protocol 自动重发请求协议
automatic resource deployment 自动资源部署
automatic search service 自动搜索服务
automatic segmentation 自动隔断
automatic sequential connection 自动按序连接
automatic service deployment 业务自动部署
automatic sounding 自动探通术
automatic switched optical network (ASON) 自动交换光网络
automatic switched transport network (ASTN) 自动交换传送网
autonomous agent for intrusion detection (AAFID) 自治代理入侵检测系统
autonomous agent network 自治代理网络
autonomous artificial neural network 自治人工神经网络
autonomous artificial neural network algorithm 自治人工神经网络算法
autonomous control system 自治控制系统
autonomous development system 自治开发系统
autonomous management system 自治管理系统
autonomous packet switching 自治信息包交换
autonomous switching 自主交换
autonomous system (AS) 自治系统
autonomous system boundary router (ASBR) 自治系统边界路由器
autonomous system number (ASN) 自治系统号
auto-partitioning 自动分隔
autoreconfiguration 自动重构
auto removal 自动拆除
autoresponder 自动响应程序
auto search 自动搜索
auto-switch default gateway 自动切换预设网关
auxiliary control network 辅助控制网络
auxiliary network address 辅助网络地址
auxiliary route 辅助路由
avatar 网上化身,网上头像
available bandwidth 可用带宽
available bit rate (ABR) 可用位速[比特]率

available bit rate service　可用位速率业务
available capacity of a network　有效网络容量
average data transfer rate　平均数据传送(速)率
average information content　平均信息量
average path length　平均路径长度
AVF (active virtual forwarder)　活跃虚拟转发器
AVG (active virtual gateway)　活跃虚拟网关
awareness network　认识网络,知晓网络
awareness of network honesty　网络诚信意识
away from keyboard (AFK)　离开键盘,暂时离开
AW-CDM (adaptive weighted code division multiplex)　自适应加权码分复用
AXP (application exchange platform)　应用交换平台
A2A (any to any)　任何到任何

# B

B (byte) 字节
b (bit) 位
back acknowledge 逆向确认
backbone communications network 主干通信网
backbone network 主干网
backbone ring 中枢环
backbone router 主干路由器
backbone site 主干节点
backbone transmission network 主干传输网
backdoor route 后门路由
back-end network 后端网
background authentication 后台认证
back link 反向链接,反链
back off 退避
backoff algorithm 退避算法
backoff counter 退避计数器
backoff interval 退避间隔
backoff time 退避时间
back orifice (BO) BO 黑客
back orifice 2000 (BO2K) BO2K 黑客工具
back-to-back gateway 背对背网关
backtracking 回溯法
backtracking control strategy 回溯式控制策略
backtracking point 回溯点

backup designated router (BDR) 备份指定路由器
backup domain controller (BDC) 备份域控制器
backup for disaster of information service system 信息服务业务系统灾难备份
backup path 后备路径
backup reverse address resolution protocol (RARP) server 备份反向地址解析协议服务器
backup reverse path 备份逆向路径
backup server 备份服务器
backup virtual path 后备虚拟路径
backward chaining 反向链接
backward channel 反向信道
backward channel capacity 反向信道容量
backward channel carrier detector 反向通道载波检测器
backward channel ready 反向通道就绪
backward channel received data (BRD) 反向通道接收数据
backward channel received line signal detector 反向通道接收线路信号检测器
backward explicit congestion notification

（BECN） 后[反]向显式拥塞通告
backward LAN channel 反向局域网信道
backward learning 反向学习
backward reasoning 逆向推理
backward search algorithm 反向搜索算法
backward setup 反向建立
backward signal 反向信号
backward supervision 反向监控
BACP（bandwidth allocation control protocol） 带宽分配控制协议
bad incoming packets 到达的坏包
bad logical connection number 错误逻辑连接数
bad network behavior 不良网络行为
Baidu index 百度指数
balanced amplitude modulation 平衡幅度调制
balanced binary tree 平衡二叉树
balanced configuration 平衡配置
balanced data link 平衡数据链路
balanced factor of node 节点的平衡因子
balanced load 平衡负载
balanced modulation 平衡调制
balanced routing 平衡路由选择
balanced transmission 平衡传输
balanced transmission line 平衡传输线
balanced unbalanced（balun） 平衡非平衡适配器
balun（balanced unbalanced） 平衡非平衡适配器
BAN（body area network） 体域网
band compaction 频带压缩

band pass 带通，通频带
bandpass signal 带通信号
bandpass signal reconstruction 带通信号重建
bandpass signal sampling 带通信号采样
bandpass wavelength division multiplexer（BWDM） 带通波分复用器
band selective repeater 选带直放站
band splitter 带宽切分器
band splitting 带分离，分频
bandwidth（BW） 带宽
bandwidth allocation control protocol（BACP） 带宽分配控制协议
bandwidth asymmetric network 带宽非对称网络
bandwidth compression 带宽压缩
bandwidth distance factor（BWDF） 带宽距离因数
bandwidth-distance product 带宽距离积
bandwidth limited 带宽受限制的
bandwidth-limited operation 带宽限制工作
bandwidth management 带宽管理
bandwidth on demand（BOD） 按需带宽
bandwidth range 带宽范围
bandwidth reservation 带宽预留
bandwidth testing 带宽测试
bandwidth throttling 带宽限制
banner 横幅广告
banner blindness 广告盲区
BAS（broadband access server） 宽带接入服务器

baseband 基带
baseband assignment 基带分配
baseband channel 基带信道
baseband frequency hopping (BFH) 基带跳频
baseband LAN 基带局域网
baseband local network 基带局部网
baseband mode 基带模式
baseband modem 基带调制解调器
baseband signal 基带信号
baseband transmission 基带传输
baseband transmission system 基带传输系统
base bandwidth 基本带宽
based on keywords search 基于关键字的搜索
based on node XML index 基于节点的XML索引
based on path XML index 基于路径的XML索引
based on prefix index 基于前缀的索引
base station controller (BSC) 基站控制器
base station identity code (BSIC) 基站识别码
basic access rate 基本访问速率
basic access signaling rate 基本接入信令速率
basic bit-map method 基本位图方法
basic encoding rules (BER) 基本编码规则
basic group 基础基群
basic information unit (BIU) 基本信息单元
basic mode control procedures (BMCP) 基本型控制规程
basic mode link control (BMLC) 基本型链路控制
basic network utilities (BNU) 基本网络实用程序
basic object adapter (BOA) 基本对象适配器
basic rate 基本速率
basic rate access (BRA) 基本速率接入
basic rate access interface 基本速率接入接口
basic rate access network termination 基本速率接入网终端
basic rate interface (BRI) 基本速率接口
basic service element (BSE) 基本业务要素
basic service 基本业务
basic service set (BSS) 基本服务群
basic serving arrangement (BSA) 基本服务配置
basis station (BS) 无线基地站
basic telecommunication access method (BTAM) 基本远程通信存取方法
basic telecommunication service 基础电信业务
basic transmission header (BTH) 基本传输标题
basic transmission unit (BTU) 基本传输单位
bastion host 桥头堡主机,堡垒机
batched EDI 批量电子数据交换
BATI (broadcast access termination identifier) 广播接入终端标识
baud rate 波特率

BBC (broadband bearer capability) 宽带承载能力
BBDLC (broadband digital loop carrier) 宽带数字环路运营商
BBIAB (be back in a bit) 马上就回
BBL (be back later) 稍后便回
BBS (bulletin board system) 电子公告牌系统
BBS (basic service set) 基本服务群
BBS (be back soon) 不久回来
.bbs 电子公告板系统文件名后缀
BCC (block-check character) 块校验字符
BCC (blind carbon copy) 暗送（邮件）
BCCC (Blockchain Collaborative Consortium) 区块链合作联盟
BCCH (broadcast control channel) 广播控制信道
BCH (broadcast channel) 广播信道
B channel (bearer channel) 承载信道，B通[信]道
BCI (bit count integrity) 比特计数完整性
BCN (beacon) 信标
BCNU (be seeing you) 正在观察
BCOB (broadband connection oriented bearer) 宽带面向连接的承载器
BCOB-A (bearer class A) 承载类型 A
BCOB-C (bearer class C) 承载类型 C
BCOB-X (bearer class X) 承载类型 X
BCU (block control unit) 块控制单元
BDC (backup domain controller) 备份域控制器
BDCS (business data center system) 业务数据中心系统

BDI (business data interchange) 商业数据交换
BDPSK (binary differential phase shift keying) 二相差分相移键控
BE (best effort) 尽力服务
beacon (BCN) 信标
beacon interval 信标间隔
beaconing station 信标站
beaconing terminal 信标终端
beacon message 信标信息
beam transport network 束流传输网络通道
bearer channel (B channel) 承载信道，B通道
bearer class A (BCOB-A) 承载类型 A
bearer class C (BCOB-C) 承载类型 C
bearer class X (BCOB-X) 承载类型 X
bearer independent call control (BICC) protocol 与承载无关呼叫控制协议
bear service (BS) 承载业务
be back in a bit (BBIAB) 马上就回
BECN (backward explicit congestion notification) 后[反]向显式拥塞通告
BECN cell BECN 信元
BED (bus extension driver) 总线扩展驱动器（卡）
BEEP (blocks extensible exchange protocol) 块可扩展交换协议
beginning of message (BOM) 消息起始
Bellman-Ford distance-vector routing algorithm 贝尔曼-福特距离向量路由选择算法
benign software 良性软件

benign viruses 良性病毒
BER（bit error rate） 位误码率，误比特率
BER（block error probability） 块差错概率
BER（basic encoding rules） 基本编码规则
BER（bus extension receiver） 总线扩展接收器（卡）
Berkeley Internet name domain（BIND） 伯克利因特网名字域
Berkeley software distribution（BSD） 伯克利软件分发
Berkeley software distribution license 伯克利软件分发许可证
be seeing you（BCNU） 正在观察
best effort（BE） 尽力服务
best-effort delivery 尽力递送
best first search 最佳优先搜索
between-the-lines entry 线间侵入
BF（brute force） 蛮力，强攻击
BFD（bidirectional forwarding detection） 双向转发检测
BFH（baseband frequency hopping） 基带跳频
BFSK（binary frequency shift keying） 二进制频移键控
BG（border gateway） 边界网关
BGMP（border gateway multicast protocol） 边界网关组播协议
BGP（border gateway protocol） 边界网关协议
BGRAN（broadband generalized radio access network） 宽带通用无线接入网络

BHO（browser helper object） 浏览器辅助对象
bias network 偏置网络
BICC（bearer independent call control） 与承载无关呼叫控制
B-ICI（B-ISDN inter-carrier interface） 宽带 ISDN 载体间接口
B-ICI SAAL（B-ICI signaling ATM adaptation layer） B-ICI 信令 ATM 适配层
B-ICI signaling ATM adaptation layer（B-ICI SAAL） B-ICI 信令 ATM 适配层
bicycle-sharing 共享单车
bidding 招标
bidirectional asymmetry 双向不对称性
bidirectional forwarding detection（BFD） 双向转发检测
bidirectional line switch ring（BLSR） 双向线路交换环
bidirectional ring 双向环
bidirectional shared tree（DST） 双向共享树
bidirectional signal 双向信号
bidirectional symmetry 双向对称性
bidirectional transmission 双向传输
biduplexed system 两双工系统
bidwitkey 悬赏型威客
bifurcated routing 分支路由选择
big data 大数据
BIGA（bus interface gate array） 总线接口门阵列
bike sharing 分享单车
bilateral control 双边控制

bilateral network  双向网络
bilateral synchronization  双向同步
binary baud rate  二进制波特率
binary channel  二进制信道
binary differential phase shift keying (BDPSK)  二相差分相移键控
binary exponential backoff  二进制指数退避
binary exponential backoff algorithm  二元指数退避算法
binary frequency shift keying (BFSK)  二进制频移键控
binary N-dimensional cube  二叉N维超立方体
binary newsgroup  二进制新闻组
binary phase shift keying (BPSK)  二相相移键控
binary runtime environment for wireless (BREW)  无线二进制运行环境
binary serial signaling rate  二进制串行信令传输速率
binary symmetric  二进制对称性
binary symmetric channel  二进制对称通道
binary synchronous communication (BSC)  二进制同步通信
binary synchronous communication protocol (BISYNC)  二进制同步通信协议
BIND (Berkeley Internet name domain)  伯克利因特网名字域
biphase coding  二相编码
biphase level code  二相电平编码
biphase mark coding  二相传号编码
biphase signaling  二相信号

biphase space coding  二相空号编码
BIP-ISDN (broadband, intelligent and personalized ISDN)  宽带化、智能化和个人化的综合业务数字网
bipolar coding  双极编码
bipolar NRZ code  双极性非归零码
bipolar NRZ signal  双极性非归零信号
bipolar with eight-zero substitution (B8ZS)  八零置换双极性代码
bipolar with six-zero substitution (B6ZS)  六零置换双极性代码
bipolar with three-zero substitution (B3ZS)  三零置换双极性代码
BIR (buffer insert ring)  缓存插入环
BISDN (broadband ISDN)  宽带综合业务数据网
BISDN inter-carrier interface (B-ICI)  宽带ISDN载体间接口
BISUP (broadband ISDN user's part)  宽带ISDN用户部分
BISYNC (binary synchronous communication protocol)  二进制同步通信协议
bisynchronous transmission  双向同步传输
bit bus  位总线
bit by bit asynchronous operation  逐步异步操作
Bitcoin (BTC)  比特币
Bitcoin exchange  比特币交易
Bitcoin Foundation  比特币基金会
Bitcoin security  比特币安全
bit count integrity (BCI)  比特计数完整性
biternary transmission  双三进制传输

bit error rate (BER)　位误码率
bit insertion　位插入
bit interleaved parity (BIP)　位交叉校验
bit interleaving　位交叉
bit interval　位间隔
bit inversion　位反转
bit map protocol　位图协议
bit misdelivery probability　比特误传概率
bit misdelivery ratio (BMR)　比特误传率
bit-oriented　面向位的
bit-oriented procedure　面向位的规程
bit-oriented protocol　面向位的协议
bit parallel transmission　位并行传输
bit physical length　位实际长度
bit pipe　比特管道
bit rate length product (BRLP)　比特率长度乘积
bit rate length product limited operation　比特率长度积受限运行
bit robbing　位劫取
bit sequence independence　比特序列独立
bit serial　位串行
bit serial transmission　位串行传输
bit slip　比特丢失
bits per second (b/s, bps)　位/秒,每秒位
bit stream　位流
bit stream transmission　位流传输
bit stuffing　位填充
bit synchronous operation (BSO)　位同步操作

bit torrent (BT)　位[比特]流
bit transparent channel　位透明信道
BIU (basic information unit)　基本信息单元
BIU (bus interface unit)　总线接口部件
BIU segment　基本信息单元段
.biz　商业网站的域名
biz. newsgroup　biz 商业新闻组
black hat Hacker　黑帽黑客
black hole　黑洞
black hole router　黑洞路由
black signal　黑信号
blind carbon copy (BCC)　隐蔽副本,盲拷贝
blind signature　盲签名
Blockchain　区块链
Blockchain capital　区块链资本
Blockchain Collaborative Consortium (BCCC)　区块链合作联盟
Blockchain nodes　区块链节点
Blockchain revolution　区块链革命
Blockchain technology　区块链技术
block check　块校验
block check character (BCC)　块校验字符
block control unit (BCU)　块控制单元
block correction efficiency factor　块校正效率因数
block efficiency　码组有效率
block error probability (BEP)　块差错概率
block error ratio (BER)　块差错率
blocking network　阻塞网络
block length (BL)　块[字组]长度

block misdelivery ratio (BMR) 码组误传率
block mode 块[分组]方式
block oriented network simulation (BONS) 面向块的网络仿真器
block parity 块奇偶校验
block rate efficiency 码组[块]速率效率
block redundancy check (BRC) 块冗余校验
block sequencing 数据块定序
blocks extensible exchange protocol (BEEP) 块可扩展交换协议
block transfer attempt 码组传送尝试
block transfer rate (BTR) 块传送速率
blog (Web log) 网(络日)志
blogger (weblogger) 博客
blog marketing 博客营销
blogosphere 博客圈
blogroll 博客链接
blog service provider (BSP) 博客服务提供商
BLP (block loss probability) 码组[块]损失概率
BLSR (bidirectional line switch ring) 双向线路交换环
BLU (basic link unit) 基本链路单位
blue book of syberspace security 网络空间安全蓝皮书
bluesnarfing 蓝牙窃用
bluetooth 蓝牙技术
bluetooth clock 蓝牙时钟
bluetooth device address (BDA) 蓝牙设备地址
bluetooth gateway 蓝牙网关

bluetooth mouse 蓝牙鼠标
bluetooth piconet 蓝牙微微网
bluetooth scatternet 蓝牙散射网
bluetooth service type 蓝牙服务类型
bluetooth Telephone 蓝牙电话
BMAN (broadband metropolitan area network) 宽带城域网
broadcast and multicast control (BMC) 广播组播控制
BMCP (basic mode control procedures) 基本型控制规程
BMI (bus memory interface) 总线内存接口
BMLC (basic mode link control) 基本型链路控制
BMR (block misdelivery ratio) 码组误传率
BMR (bit misdelivery ratio) 比特误传率
BN (bridge number) 桥号
BNC (British Naval Connector) connector BNC连接器
BNC plugs and jacks BNC插头插座
BNN (boundary network node) 边界网络节点
BNU (basic networking utilities) 基本网络实用程序
BOD (bandwidth on demand) 按需带宽
body area sensor network 体域传感器网络
body area network (BAN) 体域网
body piconet 人体微微网
BONS (block oriented network simulation) 面向块的网络仿真器

bookmark 书签
bookmark file 书签文件
bookmark list 书签(列)表
BootP (bootstrap protocol) 引导协议
BootP message format 引导协议报文格式
BootP retransmission policy 引导协议重发策略
BootP server/client 引导协议服务器/客户机
BootP two-step bootstrap procedure 引导协议两步引导过程
BootP vendor-specific field 引导协议厂家专用区段
boot ROM 引导只读存储器
bootstrap protocol (BootP) 引导协议
BOP (byte-oriented protocol) 面向字节的协议
border gateway (BG) 边界网关
border gateway multicast protocol (BGMP) 边界网关组播协议
border gateway protocol (BGP) 边界网关协议
border node 边界节点
border peer 边界对等者
border protection gateway 边界防护网关
border router (BR) 边界路由器
border router marking 边界路由器标记
BOSS (business operation support system) 业务运营支撑系统
bot (robot) 机器人,网上机器人
Botnet 僵尸网络
bounced mail 反弹邮件

bounced message 弹回消息
bounce rate 跳出率
boundary function 边界功能
boundary network node (BNN) 边界网络节点
boundary node 边界节点
bozo 傻瓜,怪人
bozo filter 傻瓜过滤器
BO2K (back orifice 2000) BO2K黑客工具
BPCF (broadband policing control function) 宽带网策略控制功能
BPDU (bridge protocol data unit) 网桥协议数据单元
BPON (broadband passive optical network) 宽带无源光网(络)
BPP (bridge port pair) 桥端口对
BPSK (binary phase shift keying) 二相相移键控
BR (border router) 边界路由器
BRA (basic rate access) 基本速率接入
bracket protocol 括号协议
brain dump 信息垃圾
brainware 脑件
BRAN (broadband radio access network) 宽带无线接入网
branch network 分支网络
BRAP (broadcast recognition with alternating priorities) 交替优先权广播识别
BRAS (broadband remote access server) 宽带远程接入服务器
BRB (be right back) 再见,马上回来
BRD (backward channel received data) 反向通道接收数据

brevity code 简缩码
brevity list 简码表
BREW（binary runtime environment for wireless） 无线二进制运行环境
BRI（basic rate interface） 基本速率接口
bridge 网桥
bridged call 桥接呼叫
bridged local area network 桥接局域网
bridged network 桥接网络
bridge forwarding 网桥转发
bridge group 网桥组
bridge identifier 网桥标识符
bridge number（BN） 桥号
bridge pages 桥页
bridge port 网桥端口
bridge port pair（BPP） 桥端口对
bridge protocol data unit（BPDU） 网桥协议数据单元
bridge router 桥由器
bridge static filtering 网桥静态过滤
bridge the digital divide 消除数字鸿沟
bridgeware 桥件
bridging 桥接
bridging loop 桥接环路
brief name 简名
bring your own device（BYOD） 自带设备
British Naval Connector（BNC） BNC连接器
BRLP（bit-rate length product） 比特率长度乘积
broadband 宽（频）带
broadband access network 宽带接入网

broadband access point 宽带接入点
broadband access server（BAS） 宽带接入服务器
broadband access unit 宽带接入设备
broadband absorption 宽带吸收
broadband bearer capability（BBC） 宽带承载能力
broadband bearer connection control 宽带承载连接控制
broadband channel 宽带信道，宽带频道
broadband coaxial system 宽带同轴电缆系统
broadband code division multiple access 宽带码分多址接入
broadband communication channel 宽带通信信道
broadband communication gateway 宽带通信网关
broadband communication network 宽带通信网
broadband connectionless data service 宽带无连接数据服务
broadband connection oriented bearer（BCOB） 面向承载的宽带连接
broadband convergence network 宽带融合网络
broadband customer premises network 宽带用户驻地网
broadband data center 宽带数据中心
broadband data channel 宽带数据通道
broadband data network 宽带数据网
broadband detection 宽带检测
broadband device 宽带设备
broadband digital cross connect system

宽带数字交叉连接系统
broadband digital loop carrier (BBDLC) 宽带数字环路运营商
broadband digital subscriber line 宽带数字用户线
broadband disturbance 宽带干扰
broadband exchange services 宽带交换业务
broadband exchange termination 宽带交换终端
broadband firewall 宽带防火墙
broadband gateway 宽带网关
broadband generalized radio access network (BGRAN) 宽带通用无线接入网络
broadband high layer information 宽带高层信息
broadband integrated distribution star 宽带综合分布式星形结构
broadband integrated distribution star network 宽带综合分布星形网
broadband integrated fiber optic network 宽带综合光纤网
broadband integrated service digital network (BISDN) 宽带综合业务数字网
broadband integrated service digital network user part 宽带综合业务数字网用户部分
broadband, intelligent and personalized ISDN (BIP-ISDN) 宽带化、智能化和个人化的综合业务数字网
broadband intelligent network 宽带智能网
broadband inter carrier interfaces 宽带互载载波间接口
broadband IP network 宽带 IP 网
broadband ISDN user's part (BISUP) 宽带 ISDN 用户部分
broadband LAN (broadband local area network) 宽带局域网
broadband local exchange carrier 宽带本地交换通信公司
broadband local network 宽带局域网
broadband local network technology 宽带局部网技术
broadband local switch 宽带市话交换
broadband lower layer information 宽带低层信息
broadband media gateway 宽带媒体网关
broadband metropolitan area network (BMAN) 宽带城域网
broadband mobile network 宽带移动网络
broadband modem 宽带调制解调器
broadband multimedia 宽带多媒体
broadband multimedia communications 宽带多媒体通信
broadband multimedia communication network 宽带多媒体通信网
broadband multimedia satellite system 宽带多媒体卫星系统
broadband multimedia service 宽带多媒体业务
broadband multimedia service system 宽带多媒体服务系统
broadband multi user communications 宽带多用户通信
broadband network 宽带网(络)

broadband network access point 宽带网络接入点
broadband network services 宽带网络服务
broadband network termination 宽带网络终端
broadband optical network terminal 宽带光网络终端
broadband packet exchange 宽带分组交换
broadband passive optical network (BPON) 宽带无源光网(络)
broadband personal communications services 宽带个人通信业务
broadband policing control function (BPCF) 宽带网策略控制功能
broadband radio access network (BRAN) 宽带无线接入网
broadband radio communication network 宽带无线通信网
broadband random vibration 宽带随机振动
broadband remote access server (BRAS) 宽带远程接入服务器
broadband remote switch unit 宽带远端交换单元
broadband router 宽带路由器
broadband satellite multimedia 宽带卫星多媒体
broadband service access point 宽带业务接入点
broadband service node 宽带业务节点
broadband signaling 宽带信令
broadband signaling network 宽带信令网络
broadband switching network 宽带交换网
broadband system 宽带系统
broadband telecommunications architecture 宽带电信体系结构
broadband terminal adapter 宽带终端适配器
broadband terminal equipment (B-TE) 宽带终端设备
broadband transfer mode (BTM) 宽带传送模式
broadband transmission 宽带传输
broadband T reference point 宽带网T参考点
broadband user network interface 宽带用户网络接口
broadband wireless access (BWA) 宽带无线接入
broadband wireless local loop 宽带无线本地环路
broadband wireless IP network 宽带无线IP网络
broadband wireless multimedia (BWM) 宽带无线多媒体
broadband wireless multimedia network 宽带无线多媒体网络
broadband wireless network 宽带无线网络
broadband wireless service center 宽带无线业务中心
broadcast 广播
broadcast access 广播接入
broadcast access key 广播接入密钥
broadcast access termination 广播接入终端

broadcast access termination identifier (BATI) 广播接入终端标识
broadcast address 广播地址
broadcast and multicast control (BMC) 广播组播控制
broadcast and select optical add drop multiplexer (BS-OADM) 广播与选择型光分插复用器
broadcast and unknown server (BUS) 广播和未知服务
broadcast area 广播区域[覆盖区]
broadcast authorization control protocol 广播授权控制协议
broadcast box 广播式信箱
broadcast channel (BCH) 广播信道
broadcast communication network 广播通信网
broadcast control channel (BCCH) 广播控制信道
broadcast control unit 广播控制单元，广播控制设备
broadcast data packet grouping protocol 广播数据包分组协议
broadcast domain 广播区域
broadcast fault tolerant routing algorithm 广播容错路由算法
broadcast feedback channel flow control 广播反馈信道流控制
broadcast file transfer protocol 广播文件传送协议
broadcast forwarding 广播转发
broadcasting to subnets 子网广播
broadcast interference 广播干扰
broadcast latency 广播延迟
broadcast mail 广播邮件

broadcast multiple access 广播多路接入
broadcast multiplexed analog component 广播多路复用模拟组件
broadcast network 广播网络
broadcast operation control 广播操作控制
broadcast packet 广播分组
broadcast protocol 广播协议
broadcast recognition access method 广播识别存取法
broadcast recognition access method protocol 广播识别存取法协议
broadcast recognition with alternating priorities (BRAP) 交替优先权广播识别
broadcast routing 广播路由选择
broadcast satellite 广播卫星
broadcast satellite service 广播卫星业务
broadcast search 广播搜索
broadcast signaling virtual channel 广播信令虚拟信道
broadcast signaling virtual channel identifier 广播信令虚拟信道标识符
broadcast storm 广播风暴
broadcast subnet 广播子网
broadcast technical standard 广播技术标准
Broadcast Television Systems Committee (BTSC) （美国）广播电视系统委员会
broadcast topology 广播拓扑
broadcast traffic 广播量

broadcast transfer control protocol 广播传送控制协议

broadcast transmission 广播传输

broadcatching 广捕,在因特网上用RSS(简易信息聚合)共享应用格式下载信息

broken link 断链

broker 中介器

brouter 网桥路由器,桥路器

browser 浏览器

BROWSER (browsing on line with selective retrieval) 带有选择性检索的联机浏览系统

browser cache 浏览器高速缓存

browser helper object (BHO) 浏览器辅助对象

browser hijacker 浏览器劫持

browser optimized content model (BOCM) 浏览器优化内容模式

browser/server architecture 浏览器/服务器体系结构

browsing on line with selective retrieval (BROWSER) 带有选择性检索的联机浏览系统

brute force (BF) 蛮力,强攻击

BS (bear service) 承载业务

BS (basis station) 无线基地站

BSA (basic serving arrangement) 基本服务配置

BSC (binary synchronous communication) 二进制同步通信

BSC (base station controller) 基站控制器

BSCA (binary synchronous communication adapter) 二进制同步通信适配器

BSIC (base station identity code) 基站识别码

BSNS (business social network site) 商务社交网站

BSO (bit synchronous operation) 位同步操作

BS-OADM (broadcast and select optical add drop multiplexer) 广播与选择型光分插复用器

BSP (blog service provider) 博客服务提供商

BSS (basic service set) 基本服务群

BT (burst tolerance) 突发容限

BTAM (basic telecommunication access method) 基本远程通信存取方法

BTC (Bitcoin) 比特币

B-TE (broadband terminal equipment) 宽带终端设备

BTH (basic transmission header) 基本传输标题

BTM (broadband transfer mode) 宽带传送模式

B to B (business to business) 企业对企业

B to C (business to customer) 企业对消费者

B to G (business to government) 企业对政府

BTR (block transfer rate) 块传送速率

BTSC (Broadcast Television Systems Committee) (美国)广播电视系统委员会

BTU (basic transmission unit) 基本传输单位

btw (by the way) 顺便提一下

buffer depletion 缓冲区耗尽
buffered asynchronous communication interface 带缓冲的异步通信接口
buffered leaky bucket algorithm 缓冲漏桶算法
buffered repeater 缓冲中继器
buffered transfer 带缓冲传输
buffer forwarding 缓存转发
buffering exchange 缓冲交换
buffer insert ring (BIR) 缓存插入环
buffer overflow attack 缓冲区溢出攻击
buffer prefix 缓冲前缀(区)
building block service 模块构建服务
building Cisco multi layer switched network 组建思科多层交换网络
building Cisco remote access network 组建思科远程接入网络
building firewalls 模块防火墙
building Internet firewalls 模块因特网防火墙
bulk E-mail 大量电子邮件,垃圾邮件
bulk redundancy 大容量冗余
bulletin board service 电子公告牌网络服务
bulletin board system (BBS) 电子公告牌系统
bump 顶
bunched frame structure 集中式帧结构
bunched optical fiber 绞合多芯光纤
bureau signaling 局间信令
burst communication 突发通信
burst error 突发差错
burst error detecting code 突发检错码

burst error detection 突发错误检测
burst isochronous 等时脉冲串(传输)
burst isochronous transmission 突发等时传输
burst mode 突发方式
burst mode communication system 突发通信系统
burst mode optical communication 突发式光通信
burst switched network 突发交换网络,成组交换网络
burst switching 突发交换
burst tolerance (BT) 突发容限
burst transmission 突发传输
BUS (broadcast and unknown server) 广播和未知服务器
bus access latency 总线访问延时
bus acquisition latency 总线获取等待时间
bus addressing 总线寻址
bus allocation 总线分配
bus and tag channel 总线和标记信道
bus arbitration 总线判优(仲裁)
bus commander 总线命令器
bus concurrency 总线并发
bus contention 总线争用
bus control 总线控制
bus controller 总线控制器
bus cycle 总线周期
bus driver 总线驱动器
bus enumerator 总线枚举器
bus error traps 总线出错中断
bus extender 总线扩展槽
bus extension driver (BED) 总线扩展驱动器(卡)

bus extension receiver (BER)　总线扩展接收器(卡)
bus family　总线簇
bus grant acknowledge　总线允许确认信号
business blog　商业博客
business communication system　商务通信系统
business communication terminal　商务通信终端
business customer premises network　商业用户驻地网
business data center system (BDCS)　业务数据中心系统
business data interchange (BDI)　商业数据交换
business information communication　商业信息通信
business information network　商业信息网
business information website　商务信息网站
business intelligent management system (BIMS)　商务智能管理系统
business marketing website　企业营销网站
business operation support system (BOSS)　业务运营支撑系统
business process execution language (BPEL)　业务流程执行语言
business social network site (BSNS)　商务社交网站
business to business (B to B, B2B)　企业对企业
business to customer (B to C, B2C)　企业对消费者
business to government (B to G, B2G)　企业对政府
business website　企业网站
business website development　商业网站开发
bus interface gate array (BIGA)　总线接口门阵列
bus interface unit (BIU)　总线接口部件
bus interlocked communication　总线互锁通信
bus interrupt　总线中断
bus local network　总线局部网络
bus lock　总线封锁
bus loop network　总线环形网络
bus master　总线主控器
bus master mode　总线主控方式
bus matrix switch　总线矩阵交换机
bus memory interface (BMI)　总线内存接口
bus message　总线报文
bus mode　总线方式
bus multiplexing　总线多路传输
bus network　总线网(络)
bus network extension　总线网络扩展
bus optical network　总线光网络
bus organized structure　总线式结构
bus organized system　总线式系统
bus oriented multiprocessor architecture　基于总线的多处理机系统结构
bus parking　总线停泊
bus polling protocol　总线查询协议
bus priority structure　总线优先结构
bus protocol　总线协议

bus quiet signal 总线安静信号
bus request 总线请求
bus responder 总线响应器
bus segment 总线区域
bus slave 总线从设备
bus slave mode 总线从控方式
bus standard 总线标准
bus status bit 总线状态位
bus switch 总线交换机,总线开关
bus system 总线系统
bus tenure 总线占用期
bus termination 总线端接法
bus timing 总线定时
bus topology 总线拓扑
bus topology control network 总线拓扑控制网络
bus topology network 总线拓扑网络
bus transaction 总线事务处理
busy channel 占用信道,忙信道
busy circuit 占用电路,忙电路
busy condition 占线状态
busy hour 忙时,高峰时
button advertising 按钮广告
buy online 网上购物
BWA (broadband wireless access) 宽带无线接入
BWDM (bandpass wavelength division multiplexer) 带通波分复用器
BWM (broadband wireless multimedia) 宽带无线多媒体
BYOD (bring your own device) 自带设备

bypass 旁路
bypass channels 旁路信道
bypass device 旁路设备
bypass mode 旁路方式
bypass relay 旁路中继
byte multiplexer 字节多路转接器
byte multiplexer channel 字节多路转换通道
byte-oriented protocol (BOP) 面向字节的协议
byte oriented serial link 面向字节的串行链路[链接]
byte oriented synchronous communication protocol 面向字节的同步通信协议
byte serial transmission 字节串行传输
bytes per second 每秒字节数,字节数/秒
byte stream file 字节流文件
byte stuffing 字节填充
byte type pipe 字节型管道
B2B (business to business) 企业对企业
B2C (business to customer) 企业对消费者
B2G (business to government) 企业对政府
B3ZS (bipolar with three-zero substitution) 三零置换双极性代码
B6ZS (bipolar with six-zero substitution) 六零置换双极性代码
B7 zero code suppression B7 零码压缩
B8ZS (bipolar with eight-zero substitution) 八零置换双极性代码

# C

CA (collision avoid) 冲突避免
CA (certificate authority) 证书认证
cable access point 电缆接入点,光缆接入点
cable adapter 电缆转接件,光缆转接件
cable area network 电缆区域网,光缆区域网
cable assembly 电缆组合件,光缆组合件
cable attenuation 电缆衰减,光缆衰减
cable balancing network 电缆平衡网络
cable-based LAN 有线局域网
cable bond 电缆接头,光缆接头
cable box 电缆箱,光缆箱,分线盒
cable branch box 电缆分线箱
cable bridge 电缆桥架
cable broadband forum (CBF) 线缆宽带论坛
cable chart 电缆连接图
cable communication system 电缆通信系统,光缆通信系统
cable component 电缆元件,光缆元件
cable configuration 电缆配置,光缆配置
cable connection 电缆连接,光缆连接
cable data modem 电缆数据调制解调器
cable interface unit (CIU) 电缆接口部件
cable link 电缆链路,光缆链路
cable maintenance center 电缆维护中心
cable matcher 电缆匹配器,光缆匹配器
cable modem (CM) 电缆调制解调器
cable modem termination system (CMTS) 电缆调制解调器终端系统
cable monitor 电缆监控器,光缆监控器
cable network 电缆网(络),光缆网(络),有线网(络)
cable news network 有线新闻网
cable noise 电缆噪声
cable number 电缆号码
cable path 电缆通路
cable sender 电缆发送机,光缆发送机
cable system 电缆系统,光缆系统
cable telephone 有线电视电话
cable television network 有线电视网
cable television relay system 有线电视中继系统
cable transfer 电缆传输,电缆传送
cable type 电缆类型,光缆类型
cable wireless hybrid network 有线无

线混合网络
cache ability 高速缓存能力
cache array 高速缓冲存储器阵列
cache array routing protocol 高速缓存阵列路由选择协议
cache bandwidth （高速）缓存带宽
cache block replacement 高速缓存块替换
cache coherence 高速缓存一致性
cache coherence protocol 高速缓存一致性协议
cache configuration 高速缓存配置
cache conflict 高速缓存冲突
cache control register 高速缓存控制寄存器
cache control unit 高速缓存控制器
cache data station 高速缓存数据站
cache directory 高速缓存目录
cache farm 高速缓存场
cache hit 高速缓存命中
cache hit ratio 缓存命中率
cache instruction station 高速缓存指令站
cache memory 高速缓冲存储器
cache memory look ahead 高速缓存预测
cache memory sharing 高速缓存共享
cache miss 高速缓存缺失，缓存未命中
cache poisoning 缓存投毒
cache server 缓存服务器
caching proxy server 缓存代理服务器
CALIS (China Academic Library & Information System) 中国高等教育文献保障系统

call abandonment probability 呼叫放弃概率
call abort 呼叫异常终止
call acceptance delay 呼叫接受延迟
call accepted condition 接受呼叫状态
call accepted message 接受呼叫消息
call accepted packet 接受呼叫分组
call accepted packet 呼叫接收包
call accepted signal 接受呼叫信号
call accounting 呼叫计次
call admission control 呼叫进入控制
call agent 呼叫代理
call allocator 呼叫分配器
callback control protocol 回叫控制协议
callback function 回呼功能
callback mechanism 回叫装置
callback modem 回调调制解调器
callback security 回叫安全(件)
callback signal 回叫信号
call block 呼叫阻塞
call by call service selection 按呼叫顺序业务选择
call by far reference 远程引用调用
call by need 按需调用
call by passing 传递调用
call by quantity 按量调用
call by result 结果调用
call canceled 呼叫取消
call classification 呼叫分类
call clearing 清除呼叫
call collision 呼叫碰撞
call collision at the DTE/DCE interface 数据终端设备/数据电路端接设备处的呼叫碰撞

call collision resolution 呼叫碰撞解除
call completion ratio 呼叫接通率
call confirmation 呼叫证实
call confirmation protocol 呼叫证实协议
call confirmation signal 呼叫证实信号
call congestion ratio 呼损率
call connect 呼叫连接
call connect signal 呼叫接续信号
call control agent 呼叫控制代理
call control block 呼叫控制块
call control character 呼叫控制字符
call control manager (CCM) 拨号控制管理
call control packet 调用控制包
call control port 调用控制端口
call control procedure 呼叫控制规程
call control signal 呼叫控制信号
call deflection selection 呼叫改发选择
call deflection subscription 呼叫改发预约
called/calling channel 被呼/呼叫信道
called channel 被呼信道
called line address modified notification 被叫线路地址修改通知
calling channel 呼叫信道
calling party number (CPN) 调用团体号
calling service user 调用服务用户
call rerouting distribution (CRD) 呼叫重选路由分配
call user data (CUD) 呼叫用户数据
CAMEL (customized application for mobile enhanced logic) 移动网增强逻辑的定制应用

camgirl 视频女
CAMP (coding aware multi path routing protocol) 编码感知多路径路由协议
campus area network 园区网,校园网
campus computer network 校园计算机网
campus network 校园网络
campus subsystem 园区子系统
campus Web network 校园 Web 网络
campus wide information system (CWIS) 园区信息服务系统
CAN (corporate area network) 企业区域网
CAN (controller area network) 控制器局域网
candidate recommendation (CR) 候选推荐标准
CANET (China Academic Network) 中国学术网
CAP (contention access period) 争用接入周期
CAPM (carrierless amplitude and phase modulation) 无载波幅相调制
CAPWAP (control and provisioning of wireless access points) 无线接入点控制和配置
car sharing 共享汽车
carrier channel 载波信道
carrier communication 载波通信
carrier communication network 载波通信网络
carrier Ethernet (CE) 电信级以太网
carrier frequency channel 载频信道
carrier frequency channel break 载频

信道阻断
carrier frequency signal transmission 载波频率信号传输
carrier identification parameter (CIP) 载体标识参数
carrierless amplitude and phase modulation (CAPM) 无载波幅相调制
carrier network 承载网络
carrier sense 载波监听
carrier sense multiple access (CSMA) 载波监听多路访问
carrier sense multiple access/collision avoid (CSMA/CA) 载波监听多路访问/冲突避免
carrier sense multiple access/collision detection (CSMA/CD) 载波监听多路访问/冲突检测
carrier sense multiple access with collision avoidance (CSMA/CA) network 具有避免冲突的载波监听多路访问网络
carrier sense multiple access with collision detection (CSMA/CD) network 载波监听多路访问/冲突检测网络
carrier sensing 信号监听
carrier signal 载波信号
carrier signaling 载波信令
carrier suppressed return zero (CSRZ) 载波抑制归零码
carrier to interference plus noise ratio (CINR) 载波干扰噪声比
carrier to interference ratio (C/I) 载频干扰比
carrier-suppressed transmission 载波抑制传输
carrier transmission 载波传输
car-sharing 共享汽车
CAS (channel associated signaling) 通道相关的信令
cascaded network 级联网络
cascade replication 瀑布式结构复制器
cascading hubs 级联集线器(组)
cascading style sheets (CSS) 层叠样式表单
cascading style sheets specification 层叠样式表单规范标准
category 1 cabling (CAT.1) 一类电缆
category 2 cabling (CAT.2) 二类电缆
category 3 cabling (CAT.3) 三类电缆
category 4 cabling (CAT.4) 四类电缆
category 5 cabling (CAT.5) 五类电缆
category 5e cabling (CAT.5e) 超五类电缆
category 6 cabling (CAT.6) 六类电缆
category 6e cabling (CAT.6e) 超六类电缆
category 7 cabling (CAT.7) 七类电缆
CAT.1 (category 1 cabling) 一类电缆
CAT.2 (category 2 cabling) 二类电缆
CAT.3 (category 3 cabling) 三类电缆
CAT.4 (category 4 cabling) 四类电缆
CAT.5 (category 5 cabling) 五类电缆
CAT.5e (category 5e cabling) 超五类电缆
CAT.6 (category 6 cabling) 六类电缆
CAT.6e (category 6e cabling) 超六类电缆
CAT.7 (category 7 cabling) 七类电缆

CAU (channel access unit) 信道接入单元

CAUCE (Coalition Against Unsolicited Commercial Email) 反未经许可商业邮件联盟

CAVE (cellular authentication and voice encryption) 蜂窝认证和语音加密算法

CBDS (connectionless broadband data service) 无连接宽带数据服务

CBF (cable broadband forum) 线缆宽带论坛

CBR (constant bit rate) 恒位速率

CBWFQ (class-based weighted fair queuing) 基于类的加权公平排队

CCCH (common control channel) 公共控制信道

CDG (CDMA development group) CDMA 开发组

CCAI (cloud computing assisted instruction) 云计算辅助教学

CCG (content charging gateway) 内容计费网关

CCH (control channels) 控制信道

CCIA (Computer and Communication iIndustry Association) 计算机和通信工业协会

CCIE (Cisco Certified Internetwork Expert) 思科认证网络专家

CCIR (Consultative Committee of International Radio) 国际无线电咨询委员会

CCIRN (Coordinating Committee for Intercontinental Research Networks) 洲际研究网络协调委员会

CCIS (common channel interoffice signaling) 公共信道局间信令

CCITT (Consultative Committee on International Telegraph and Telephone) 国际电报电话咨询委员会

CCITT channel type 国际电报电话咨询委员会(CCITT)建议的通道类型

CCITT I series recommendations CCITT 的 I 系列建议

CCITT V series recommendations CCITT 的 V 系列建议

CCITT X series recommendations CCITT 的 X 系列建议

CCN (cluster controller node) 簇控制器节点

CCNA (Cisco Certified Network Associate) 思科认证网络工程师

CCNP (Cisco Certified Network Professional) 思科认证网络专业人员

CCR (current cell rate) 当前信元速率

CCR (commitment, concurrency and recovery) 提交、并发和恢复

CCS (common channel signaling) 公共信道信令

CCS (common communication support) 公共通信支持

CCSA (China Communications Standards Association) 中国通信标准化协会

CCSN (common channel signaling network) 公共信道信令网

CCSS (common-channel signaling system) 公共信道信令系统

ccTLD (country code top level domain names) 国家代码顶级域名

CCU (communication control unit) 控

制单元
CD (carrier detect) 载波检测
CD (collision detection) 冲突检测
CDDI (copper distributed data interface) 铜线分布数据接口
CDF (channel definition format) 频道定义格式
CDM (compound document mail) 组合文档邮件
CDM (color division multiplexing) 色分多路复用
CDMA (code division multiple access) 码分多址
CDMA development group (CDG) CDMA 开发组
CDMA 2000 standard CDMA2000 标准
CDMI (cloud data management Interface) 云数据管理接口
CDN (content delivery network) 内容传递网
CDN (content distribution network) 内容分发网络
CDN (Chinese domain name) 中文域名
CDNM session (cross-domain network manager session) 跨域网络管理器会话
CDP (Cisco discovery protocol) 思科发现协议
CDPD (cellular digital packet data) 蜂窝数字式分组数据交换
CDPSK (coherent differential phase shift keying) 相干差分相移键控
CD-ROM mirror server 只读光碟镜像服务器

CDSL (consumer digital subscriber line) 消费者数字用户线
CDV (cell delay variation) 信元时延变化
CDVT (cell delay variation tolerance) 信元时延变化容限
CE (custom edge) 用户边缘设备
CE (carrier Ethernet) 电信级以太网
CE (consumer electronic) 消费电子
CECA (China Electronic Commerce Association) 中国电子商务协会
CEFL (clusterhead election using fuzzy logic) 模糊逻辑的群首选举
CEI (comparably efficient interconnection) 同等的互连
CEI (connection endpoint identifier) 连接终止端点标识符
CEINET (China Economy Information Net) 中国经济信息网
cell delay variation (CDV) 信元时延变化
cell delay variation tolerance (CDVT) 信元时延变化容限
cell error ratio (CER) 信元出错率
cell header 信元头部
cell loss priority (CLP) 信元丢失优先级
cell loss ratio (CLR) 信元丢失率
cell mean delay time 信元平均时延
cell misinsertion rate (CMR) 信元误插率
cell neuron network 细胞神经元网络
cell on wheels (COW) 车载移动式基站
cell output switch 信元输出交换

cell rate margin (CRM)　信元率边缘
cell relay　信元中继
cell relay function (CRF)　信元中继功能
cell relay service (CRS)　信元中继服务
cells in flight (CIF)　传输中信元
cells in frame (CIF)　帧中信元
cell switching　信元交换
cells per second (CPS)　每秒信元数
cell time delay　信元时延
cell transfer delay (CTD)　信元传输延迟
cellular authentication and voice encryption (CAVE)　蜂窝认证和语音加密算法
cellular data network　蜂窝数据网
cellular data suite (CDS)　蜂窝数据套件
cellular digital packet data (CDPD)　蜂窝数字式分组数据交换
cellular memory　细胞存储器
cellular message encryption algorithm　蜂窝信息加密算法
cellular mobile communication (CMC)　蜂窝移动通信
cellular mobile communication network (CMCN)　蜂窝移动通信网
cellular mobile data network　蜂窝移动数据网
cellular mobile IP network　蜂窝移动IP网
cellular mobile network　蜂窝移动网络
cellular mobile radiotelephone system　蜂窝移动无线电话系统
cellular mobile telephone network　蜂窝移动电话网
cellular mobile telephony　蜂窝移动电话
cellular network　细胞网络
cellular network　蜂窝网络
cellular neural network　细胞神经网络
cellular neural network systems　细胞神经网络系统
cellular packet switching　蜂窝分组交换
cellular radio data network　蜂窝无线数据网络
cellular signalling network　蜂窝信号网络
cellular wireless network　蜂窝无线网络
cell variation delay tolerance (CVDT)　信元时延变动的允许范围
CELP (code excited linear prediction)　码激励线性预测（编码）
CEN (China Education Network)　中国教育网
center authentication server　中央认证服务器
center resource management (CRM)　中心资源管理
central bridge　中心网桥
central directory　中央目录
central directory server (CDS)　中央目录服务器
central exchange　集中式用户交换机
centralized adaptive routing　集中式自适应路由选择
centralized algorithm　集中式算法
centralized control network　集中控制

网络
centralized management　集中管理
centralized mini-slot packet reservation multiple access protocol　集中式微时隙分组预留多址协议
centralized network　集中式网络
centralized network architecture（CNA）集中式网络体系结构
centralized network control facility　集中式网络控制设备
centralized network management　集中式网络管理
centralized network security events management system　集中式网络安全事件管理系统
centralized operation　集中式操作
centralized routing　集中式路由选择
concentrative token control　集中式令牌控制
centralized token passing（CTP）集中式令牌传送
centralized topology　集中式拓扑结构
centralized traffic control　集中流量控制
centralled computer network　集中式计算机网络
central office local area network（COLAN）中心局域网
central office terminal（COT）局端机
central resource registration　中央资源登记
CERNET（China Education and Research Network）中国教育和科研计算机网
CERT（computer emergency response teams）计算机应急响应组
certificate authority（CA）证书认证
certificate authority center　证书认证中心
certificate hierarchy　证书层次结构
certificate trust list（CTL）证书信用［托管］表
certification revocation list（CRL）证书注销表
certified information security professional（CISP）注册信息安全专家
certified information systems security professional（CISSP）信息系统安全认证专家
certified Novell administrator　合格Novell网络管理员
certified Novell engineering　合格Novell网络工程师
certified Novell engineering professional association　合格Novell网络工程师专业协会
certified Novell instructor　合格Novell网络指导者
certified wireless networking professional（CWNP）无线网络专家认证
CFCA（China Financial Certificate Authority）中国金融认证中心
CFN（connection frame number）连接帧号
CFSK（coherent frequency shift keying）相干频移键控
CGI（common gateway interface）公共网关接口
cgi-bin（common gateway interface-binaries）cgi-bin目录,公共网关接

口目录
CGI scripts 公共网关接口脚本
CGSR (clusterhead gateway switch routing) 群首网关交换路由协议
CGWNet (China Great Wall Network) 中国长城网
chain error 链错误
chain letter 连锁信
challenge and reply 询问和应答
challenge and reply authentication 询问和应答验证
challenge handshake authentication protocol (CHAP) 询问握手认证协议
chamical markup language (CML) 化学标记语言
change direction protocol 换向协议
change-over 倒换,转换
channel access 通道访问
channel access protocol 信道接入协议
channel access unit (CAU) 信道接入单元
channel acquisition delay 信道获取延迟
channel adapter input/output supervisor 通道适配器输入输出管理程序
channel allocation 通道分配
channel allocation network 信道分配网络
channel associated signaling (CAS) 通道相关的信令
channel-attachment major node 通道连接主节点
channel borrowing scheme 借用信道
channel bypass 旁路信道
channel capacity 信道流量[传输率]

channel coding 信道编码
channel controller (CC) 通道控制器
channel definition format (CDF) 频道定义格式
channel equalization 信道均衡
channel estimation 信道估计
channel extension 信道扩展
channel group 信道群
channel group rate 信道群速率
channel hopping 信道跳跃
channel isolation 信道隔离度
channelized E1 信道化的 E1 线路
channelized T1 信道化的 T1 线路
channelizing 信道化
channel jumbo group 信道巨群
channel marketing network 渠道营销网络
channel master group 信道主群
channel network 信道网络
channel operator 通道操作者
channel power equalization 通道功率均衡
channel routing 信道路由选择
channel service unit (CSU) 通道服务单元
channel spacing 信道间隔
channel supergroup 超群信道
channel time slot 信道时隙
channel transfer rate 信道传送速率
CHAP (challenge handshake authentication protocol) 询问握手认证协议
character-oriented 面向字符
character-oriented procedure 面向字符规程
character-oriented protocol 面向字符

协议

character synchronized　字符同步

chat on line　在线聊天

chat tool　聊天工具

cheapernet　廉价网络

checking symbol　校验符号

China Academic Library & Information System (CALIS)　中国高等教育文献保障系统

China Academic Network (CANET)　中国学术网

China Communications Standards Association (CCSA)　中国通信标准化协会

ChinaDDN (China Digital Data Network)　中国数字数据网

China Digital Data Network (ChinaDDN)　中国数字数据网

China Domain Name (CDN)　中国域名

China Economy Information Net (CEINET)　中国经济信息网

China Education and Research Network (CERNET)　中国教育和科研计算机网

China Education Network (CEN)　中国教育网

China Electronic Commerce Association (CECA)　中国电子商务协会

China Financial Certificate Authority (CFCA)　中国金融认证中心

ChinaGBN (China Golden Bridge Network)　中国金桥信息网

China Golden Bridge Network (ChinaGBN)　中国金桥信息网

China Great Wall Network (CGWNet)　中国长城网

China Information Security Certification Center (ISCCC)　中国信息安全认证中心

China International Economy and Trade Net (CIETNet)　中国国际经济与商业网

China Internet Network Information Center (CNNIC)　中国互联网络信息中心

China Mobile Net (CMNet)　中国移动互联网

China National Network (CNFN)　中国国家金融网

China National Knowledge Infrastructure (CNKI)　中国知识基础设施

ChinaNet　中国公用计算机互联网

China Netcom (CNCnet)　中国网通

China Network Audio-Visual Industry Forum (CNAIF)　中国网络视听产业论坛

China Network Television (CNTV)　中国网络电视台

China next generation internet (CNGI)　中国下一代互联网

China Organizational Name Administration Center (CONAC)　（中国）政务和公益机构域名注册管理中心

ChinaPAC (China Public Packet Switched Data Network)　中国公用分组交换数据网

ChinaPSTN (China Public Switched Telephone Network)　中国公用电话网

China Public Packet Switched Data Network (ChinaPAC)　中国公用分组交换数据网

China Public Switched Telephone Network (ChinaPSTN) 中国公用电话网

ChinaSat (China Telecommunication Broadcast Satellite Corporation) 中国通信广播卫星公司

China Satellite Internet (CSNet) 中国卫星因特网

China Science and Technology Network (CSTNET) 中国科技网

China Telecommunication Broadcast Satellite Corporation (ChinaSat) 中国通信广播卫星公司,中国卫通

China Wireless Telecommunications Standard (CWTS) 中国无线通信标准

Chinese domain name (CDN) 中文域名

Chinese Science Citation Database (CSCD) 中国科学引文数据库

.chm HTML 文件名后缀,超文本标记语言文件名后缀

CHNM (simple home network management) 简单家庭网络管理

choke packet 抑制分组

CHTML (compact hypertext markup language) 压缩式超文本标记语言

CI (congestion indicator) 拥塞指示符

C/I (carrier to interference ratio) 载频干扰比

CIA (classical IP over ATM) 在 ATM 网络上运行传统 IP 业务

CICP (communication interrupt control program) 通信中断控制程序

CIDF (common intrusion detection framework) 通用入侵检测框架

CIDR (classless inter-domain router) 无类域间路由

CIF (cells in flight) 传输中信元

CIF (cells in frame) 帧中信元

CIFS (common internet file system) 公用网际文件系统

CIM (common information model) 公共信息模型

CINR (carrier to interference plus noise ratio) 载波干扰噪声比

CIP (carrier identification parameter) 载体标识参数

CIP (common indexing protocol) 通用索引协议

CIP (commerce interchange pipeline) 商务交换管道

circuit group 电路组

circuit switch data transmission service 电路交换数据传输服务

circuit switched data network (CSDN) 电路交换数据网

circuit switched digital network (CSDN) 电路交换数字网络

circuit switched gateway (CS-GW) 电路交换网关

circuit switched network (CSN) 电路交换网络

circuit-switched voice (CSV) 电路交换语音

circuit switching center (CSC) 电路交换中心

circuit switching delay 电路转换延迟

circuit switching multiplexor (CS-MUX) 电路交换多路复用器

circuit switching unit (CSU) 电路交换单元

circuit transfer mode 电路传输模式
Cisco Certified Internetwork Expert (CCIE) 思科认证网络专家
Cisco Certified Network Associate (CCNA) 思科认证网络工程师
Cisco Certified Network Professional (CCNP) 思科认证网络专业人员
Cisco discovery protocol (CDP) 思科发现协议
CIX (Commercial Internet Exchange Association) 商用因特网网际交换协会
class A IP address A 类 IP 地址
class A network A 类网络
class A network address A 类网址
class-based queuing (CBQ) 基于类的排队
class-based weighted fair queuing (CBWFQ) 基于类的加权公平排队
class B IP address B 类 IP 地址
class B network address B 类网址
class C IP address C 类 IP 地址
class C network address C 类网址
class D address D 类地址
class D network address D 类网址
class E channel E 类通道
class E IP address E 类 IP 地址
classful IP address 全类 IP 地址
classful routing protocol 有类路由协议
classical IP 经典网际协议,传统网际协议
classical IP over ATM (CIA) 在 ATM 网络上运行传统 IP 业务,在异步传输模式网络上运行传统 IP 业务
classified security protection of information service system 信息服务业务系统安全等级保护
classified security protection of instant messaging system 即时消息业务系统安全等级保护
classified security protection of the domain name registration system 域名注册系统安全等级保护
classless inter-domain routing (CIDR) 无类域间路由
classless inter-domain routing list 无类域间路由表
classless IP address 无类别 IP 地址
classless routing 无类路由
classless routing protocol 无类路由协议
class of service (CoS) 服务级(别)
class of service database 服务数据库类
CLAW (common link access for workstations) 工作站的通用链路访问
CLEC (competitive local exchange carriers) 竞争本地交换运营商
CLI (cluster link interface) 簇链路接口
click distance 单击距离
click ratio 点击率
click stream 点击流
click through 点击次数
click through ratio (CTR) 点击通过率
clickwrap agreement 点击式许可协议书
client 客户端,客户机
client access license 客户访问许可协议
client end node 客户终端节点

client error 客户端错误
client/network model 客户机/网络模式
client process 客户机进程
client program 客户机程序
client/server (C/S) 客户机/服务器
client/server architecture 客户机/服务器体系结构
client/server computing 客户机/服务器计算
client/server computing feature 客户机/服务器计算特点
client/server framework 客户机/服务器架构
client/server model 客户机/服务器模型
client/server network 客户机/服务器网络
client/server protocol 客户机/服务器协议
client/server structure 客户机/服务器结构
client/server system (CSS) 客户机/服务器系统
client-side 客户端
client-side application 客户端应用程序
client-side attacks 客户端攻击
client-side certificate 客户端认证
client-side interceptor 客户端拦截器
client-side plugins 客户端插件
client-side program 客户端程序
client signal fail (CSF) 客户端信号失效
client strategy routing 客户端策略路由

client tier 客户端层
client tier component 客户层组件
client workstation 客户工作站
CLLM (consolidated lin layer management) 统一链路层管理协议
CLNP (connectionless network protocol) 无连接网络协议
CLNS (connection less network service) 无连接网络服务
closed flow network 闭型流动网络
closed-loop transmit diversity (CLTD) 闭环发信分集
closed network 闭合网络
closed user group (CUG) 闭合用户组
cloud client computing 云客户端计算
cloud computer 云计算机
cloud computing 云计算
cloud computing assisted instruction (CCAI) 云计算辅助教学
cloud computing and developer platform 云计算与开发平台
cloud computing and information integration 云计算与信息集成
cloud computing federation 云计算联盟
cloud computing platform 云计平台
cloud computing technology 云计算技术
cloud content delivery network 云内容发布网络
cloud crowdsourcing 云众包
cloud backup and disaster recovery 云灾备,云备份和灾难恢复
cloud data center 云数据中心
cloud data management 云数据管理

cloud data management Interface (CDMI) 云数据管理接口
cloud data storage 云数据存储
cloud delivery network 云分发网络
cloud integration 云整合
cloud manager 云管理器
cloud management 云管理
cloud mobile computing 云移动计算
cloud network 云网络
cloud network engineer 云网络工程师
cloud platforms 云平台
cloud QuickPass 云闪付
Cloud Security Alliance (CSA) 云安全联盟
cloud server 云服务器
cloud software 云软件
cloud software as a service 云软件即服务
cloud storage 云存储
cloud TV 云电视
cloud systems administrator 云系统管理员
cloud Web 云网络
cloud Web hosting 云主机
CLP (command line protocol) 命令行协议
CLP (cell loss priority) 信元丢失优先级
CLR (cell loss ratio) 信元丢失率
CLT (communication line terminal) 通信线路终端
CLTD (closed-loop transmit diversity) 闭环发信分集
CLTP (connectionless transport protocol) 无连接传输协议
CLTS (connectionless transport service) 无连接传输服务
cluster 群集(站),集群
cluster attack 聚类攻击
cluster-based Web caching system 网络集群缓存系统
cluster computing 簇计算
cluster controller 簇控制器
cluster controller node (CCN) 簇控制器节点
cluster grid computing 集群网格计算
clusterhead election using fuzzy logic (CEFL) 模糊逻辑的群首选举
clusterhead gateway switch routing (CGSR) 群首网关交换路由协议
cluster interface 簇接口,群接口
cluster link interface (CLI) 簇链路接口
cluster multichannel MAC protocol (CMMP) 集群多信道 MAC 协议,集群多信道介质访问控制协议
cluster processor 聚类处理器
CM (cable modem) 电缆调制解调器
CMC (communication management configuration) 通信管理配置
CMC (cellular mobile communication) 蜂窝移动通信
CMCN (cellular mobile communication network) 蜂窝移动通信网
CMCS (computer mediated communication service) 计算机媒介通信服务系统
CMIP (common management information protocol) 公共管理信息协议
CMIP (common management interface protocol) 共同管理接口协议

CMIS (common management information service) 公共管理信息服务
CMTS (cable modem termination system) 电缆调制解调器终端系统
CML (chamical markup language) 化学标记语言
CMMP (cluster multichannel MAC protocol) 集群多信道 MAC 协议，集群多信道介质访问控制协议
CMNet (China Mobile Net) 中国移动互联网
CMNP (connection-mode network protocol) 连接(模)式网络协议
CMOT (common management information service protocol over TCP/IP) 在 TCP/IP 上的公共管理信息服务协议
CMS (content management system) 内容管理系统
CMT (connection management) 连接管理
CN (core network) 核心网
CNA (communication network architecture) 通信网络体系结构
CNA (centralized network architecture) 集中式网络体系结构
CNAIF (China Network Audio-Visual Industry Forum) 中国网络视听产业论坛
CNAP (connectionless network access protocol) 无连接网络存取协议
CNLP (connectionless network layer protocol) 无连接网络层协议
CNCERT/CC (National Computer Network Emergency Response Technical Team/Coordination Center of China) 中国国家计算机网络应急技术小组/处理协调中心
CNFN (China National Network) 中国国家金融网
CNGI (China next generation internet) 中国下一代互联网
CNKI (China National Knowledge Infrastructure) 中国知识基础设施
CNM (communication network management) 通信网络管理
CNM (congestion notification message) 拥塞通知消息
CNN (convolutional neural network) 卷积神经网络
CNNIC (China Internet Network Information Center) 中国互联网络信息中心
CNOM (Committee of Network Operation and Management) 网络营运与管理委员会
CNR (communication networking riser) 通信网络插卡
CNTV (China Network Television) 中国网络电视台
Coalition Against Unsolicited Commercial Email (CAUCE) 反未经许可商业邮件联盟
co-and contra-directional interface 同向和反向接口
coarse wavelength division multiplexing (CWDM) 稀疏波分复用
coaxial cable HSLN 同轴电缆高速局部网
co-channel 同信道
co-channel interference 同信道干扰
co-channel interference reduction factor

同频干扰降低系数

COD (connection oriented data) 面向连接的数据

codebook attack 电码本攻击

codebook excited linear prediction 码本激励线形预测编码

code channel 代码信道

code division multiple access (CDMA) 码分多址,码分多路访问

code division multiplexing (CDM) 码分复用

coded orthogonal frequency division multiplex (COFDM) 编码正交频分复用

code excited linear prediction (CELP) 码激励线性预测(编码)

code excited linear prediction (CELP) compression 编码激励线性预测压缩

code-independent data communication 码独立数据通信

code signing 代码签署

code-transparent data communication (代)码透明数据通信

coding aware multi path routing protocol (CAMP) 编码感知多路径路由协议

cognitive computing 认知计算

cognitive pilot channel (CPC) 感知导频信道

cognitive radio (CR) 认知无线电

coherent bandwidth 相干带宽

coherent communication system 相干通信系统

coherent detection 相干检波

coherent differential phase shift keying (CDPSK) 相干差分相移键控

coherent echo 相干回波

coherent frequency shift keying (CFSK) 相干频移键控

coherent multicarriers 相干多载波

coherent network 一致[相干]网络

cohesive network 内聚网络

coincidence attack 重合攻击

COLAN (central office local area network) 中心局局域网

cold fusion markup language (CFML) 冷聚变标记语言

ColdFusion server 冷聚变服务器

cold link 冷链接

collaborative filtering 协作过滤

collaborative filtering recommendation algorithms 协作过滤推荐算法

collaborative software 协作软件

collaborative virtual environments (CVE) 协作虚拟环境

collaborative virtual manufacturing (CVM) 协作虚拟制造

collaborative virtual reference service 协作虚拟咨询服务

collapsed backbone 紧缩主干网

collective routing 集群路由选择

collective routing indicator 集群路由指示符

collision 碰撞,冲突

collision avoid (CA) 冲突避免

collision avoidance 冲突预防

collision detection (CD) 冲突检测

collision detection methods 冲突检测方法

collision domain 冲突域

collision enforcement 冲突强制

collision-free protocols 无冲突协议
collision resolutions technique 冲突分解技术
collision signal 冲突信号
collision window 冲突窗口
color division multiplexing (CDM) 色分多路复用
.com 商业机构域名
combined station 组合站
command line protocol (CLP) 命令行协议
command, control, computer, and communication (C4) 指挥、控制、计算机和通信
command net 命令网络
command protocol data unit 命令协议数据单元
command terminal protocol (CTP) 命令终端协议
command translator 命令翻译机
command transmit time out 命令传送超时
commerce interchange pipeline (CIP) 商务交换管道
commerce XML (cXML) language 商用扩展置标语言
commercial communication satellite 商用通信卫星
commercial communication service 商用通信业务
Commercial Internet Exchange Association (CIX) 商用因特网网际交换协会
commercial network 商用型网
commercial online service (COLS) 商业联机业务
commitment, concurrency and recovery (CCR) 提交、并发和恢复
committed burst rate (CBR) 承诺[约定]突发速率
committed information rate (CIR) 承诺信息速率
Committee of Network Operation and Management (CNOM) 网络营运与管理委员会
commodity promotion solution (CPS) 商品推广解决方案
common abbreviations 通用缩略语
common agent technology 共用代理技术
common application layer protocol 公共应用层协议
common bonding network 公共连接网络
common carrier wide band channels 公用载波宽带信道
common channel interoffice signaling (CCIS) 公共信道局间信令
common channel signaling (CCS) 公共信道信令,公路信令
common channel signaling network (CCSN) 公共信道信令网,共路信令网
common-channel signaling system (CCSS) 公共信道信令系统,共路信令系统
common-channel signaling system No. 7 7号公共信道信令,7号共路信令
common communication network 公用通信网
common communication support (CCS) 公共通信支持
common control channel (CCCH) 公共

控制信道
common control switching arrangement (CCSA) 共用控制交换设备
common data network 公共数据网络
common data security architecture (CDSA) 公共数据安全体系结构
common gateway 公共网关,通用网关
common gateway interface (CGI) 公共网关接口
common gateway interface script 公共网关接口脚本
common indexing protocol (CIP) 通用索引协议
common information model (CIM) 公共信息模型
common internet file system (CIFS) 公用网际文件系统
common internet file system/enterprise (CIFS/E) 公用网际文件系统扩展（协议）
common intrusion detection framework (CIDF) 通用入侵检测框架
common intrusion specification language (CISL) 通用入侵规范语言
common link access for workstations (CLAW) 工作站的通用链路访问
common management information protocol (CMIP) 公共管理信息协议
common management information service (CMIS) 公共管理信息服务
common management information service definition 公共管理信息服务定义
common management information service element (CMISE) 公共管理信息服务元素
common management information service protocol over TCP/IP (CMOT) 在 TCP/IP 上的公共管理信息服务协议
common management interface protocol (CMIP) 共同管理接口协议
common network 公用网络
common object request broker architecture (CORBA) 公共对象请求代理体系结构
common open policy service (COPS) 公共开放策略服务协议
common open policy service usage for policy provisioning (COPS-PR) 公共公共策略服务的策略提供功能
common open software environment (COSE) 通用开放软件环境
common open system environment (COSE) 通用开放系统环境
common operations environment (COE) 公共操作环境
common part convergence sublayer (CPCS) 公共部件会聚子层
common part convergence sublayer-service data unit (CPCS-SDU) 公共部件会聚子层-服务数据单元
common part sublayer (CPS) 公共部分子层
common peer group 公共对等组
common physical channel identifier (CPCId) 公共物理信道标识
common programming interface 公共编程接口
common transport channel identifier (CTCId) 公共传输信道标识
common service network 公用业务

网络
common user access (CUA) architecture 公共用户访问系统结构
common user network 公用用户网
common user service 公共用户业务
Common Vulnerabilities and Exposures (CVE) 通用漏洞披露(组织)
communication access method 通信访问方法
communication automation (CA) 通信自动化
communication automation system 通信自动化系统
communication channel 通信信道
communication channel capacity 信道容量
communication channel carrying capacity 信道承载能力
communication circuit 通信电路
communication control character 通信控制字符
communication controller 通信控制器
communication controller node 通信控制器节点
communication control procedure 通信控制规程
communication control unit (CCU) 通信控制单元
communication file transfer 通信文件传送
communication intelligence channel 通信智能通道
communication interface 通信接口
communication interface layer 通信接口层

communication interface modules 通信接口模块
communication interrupt control program (CICP) 通信中断控制程序
communication line terminal (CLT) 通信线路终端
communication management configuration (CMC) 通信管理配置
communication management host 通信管理主机
communication multiplexer channel 通信多路复用信道
communication network architecture (CNA) 通信网络体系结构
communication networking riser (CNR) 通信网络插卡
communication network management (CNM) 通信网络管理
communication network processor 通信网处理机
communication network simulator 通信网模拟器
communication network topology 通信网络拓扑
communication online test system 通信联机测试系统
communication optical cable 通信光缆
communication overhead 通信开销
communication partner 通信(合作)伙伴
communication path 通信路径
communication platform 通信平台
communication port 通信端口
communication process 通信进程
communication processor 通信处理机

communication protocol 通信协议
communication private network 通信专网
communication recovery 通信恢复
communication relay base 通信中继基站
communication reliability 通信可靠性
communication request 通信请求
communication resource 通信资源
communication retransmitting station 通信转发站
communication route 通信路由
communication routing table 通信路由表
communication satellite network 通信卫星网
communication satellite space station 通信卫星空间站
communication section 通信节
communication security 通信安全
communication security monitoring 通信安全监听
communication sequential processing (CSP) 通信顺序处理
communication server 通信服务器
communication service interface 通信服务接口
communication signal interception 通信信号截获
communication signal monitoring 通信信号监听
communication simulator 通信仿真器
communication slot 通信插槽
communication society 通信协会
communication speed 通信速度

communication standard 通信标准
communication subnet 通信子网
communication transmission network 通信传输网
community computing 社区计算
compact hypertext markup language (CHTML) 压缩式超文本标记语言
comparably efficient interconnection (CEI) 同等的互连
comparison shopping 比较购物
comparison shopping search engine 比较购物搜索引擎
compatible with IPv4 address 兼容 IPv4 地址
competitive local exchange carriers (CLEC) 竞争本地交换运营商
composite bus enumerator 复合总线枚举器
compound document mail (CDM) 组合文档邮件,混合文档邮件
compound fuzzy neural network 复合型模糊神经网络
compound modularization network 复合模块化网络
compound network space 复合网络空间
compound neural network 复合神经网络
compound orthogonal neural network 复合正交神经网络
CompTIA (Computing Technology Industry Association) (美国)计算机技术工业协会
CompTIA certification CompTIA 认证,(美国)计算机技术工业协会认证

Computer and Communication iIndustry Association (CCIA) 计算机和通信工业协会
computer city 计算机城市
computer communication 计算机通信
computer communication network 计算机通信网
computer distribution network 计算机分布式网络
computer emergency response teams (CERT) 计算机应急响应组
computer emergency management system 计算机应急管理系统
computer information system 计算机信息系统
computer mediated communication service (CMCS) 计算机媒介通信服务系统
computer model for network analysis 网络分析的计算机模型
computer name 计算机名
computer network 计算机网络
computer network facilities 计算机网络设施
Computer Network Information Center (CNIC) 计算机网络信息中心
computer network system 计算机网络系统
computer-supported collaboration work (CSCW) 计算机支持的协作工作
computer telecommunication integration (CTI) 计算机电信集成
computer telephony image (CTI) 三网合一
computer telephony integration (CTI) 计算机电话集成
Computing Technology Industry Association (CompTIA) （美国）计算机技术工业协会
CONAC (China Organizational Name Administration Center) （中国）政务和公益机构域名注册管理中心
concentrated gateway 集中式网关
concentrator 集中器
concentrator device 集中器设备
conceptual integration network 概念整合网络
conceptual network 概念网络
conceptual reasoning network 概念推理网络
conceptual semantic network 概念语义网络
concurrent remirror requests 并行重镜像请求
configuration builder 配置生成器
configuration cell of network 网络结构单元
configuration exchange utility 配置交换实用程序
configuration management (CM) 配置管理
configuration network management system 配置网络管理系统
configuration network variable 可配置网络变量
configuration of network 网络配置
configuration of peer network 对等网配置
configuration parameters 配置参数
configuration register value 配置寄

存值
configuration service　配置服务
confirmation E-mail　确认电子邮件
conflict resolution strategies　冲突消解策略
conformant management entity (CME)　一致管理项目
congestion　拥塞
congestion avoidance　拥塞避免
congestion collapse　拥塞崩溃
congestion control　拥塞控制
congestion control algorithms　拥塞控制算法
congestion control mechanism　拥塞控制机制
congestion indicator (CI)　拥塞指示符
congestion control　拥塞控制
congestion management　拥塞管理
congestion notification message (CNM)　拥塞通知消息
congestion window　拥塞窗口
congestion window size　拥塞窗口大小
connect data set to line (CDSTL)　连接数据集到线路
connected network　连通网络
connection control interface (CCI)　连接控制接口
connection delay　连接延迟
connection end point (CEP)　连接(终止)端点
connection endpoint identifier (CEI)　连接终止端点标识符
connection frame number (CFN)　连接帧号
connectionless　无连接

connectionless broadband data service (CBDS)　无连接宽带数据服务
connectionless file server　无连接文件服务器
connectionless gateway　非连接网关
connectionless-mode network service　无连接式网络业务
connectionless-mode transmission　无连接式传输
connectionless network　无连接网络
connectionless network access protocol (CNAP)　无连接网络存取协议
connectionless network layer protocol (CNLP)　无连接网络层协议
connectionless network protocol (CLNP)　无连接网络协议
connectionless network service (CLNS)　无连接网络服务
connectionless node network service　无连接节点网络服务
connectionless protocol　无连接协议
connectionless service (CLS)　无连接服务
connectionless transport network　无连接传送网
connectionless transport protocol (CLTP)　无连接传输协议
connectionless transport service (CLTS)　无连接传输服务
connection management (CMT)　连接管理
connection-mode network protocol (CMNP)　连接(模)式网络协议
connection-mode network service (CMNS)　连接式网络业务

connection-mode transmission 连接（模）式传输
connection network 连接网络
connection number 连接号
connection-oriented 面向连接的
connection-oriented data (COD) 面向连接的数据
connection-oriented data transfer protocol 面向连接的数据传送协议
connection-oriented gateway 面向连接网关
connection-oriented mode 面向连接的模式
connection-oriented mode transmission 面向连接模式的传输
connection-oriented network 面向连接网络
connection-oriented network protocol (CONP) 面向连接网络协议
connection-oriented protocol 面向连接协议
connection-oriented service 面向连接的服务
connection point (CP) 连接点
connection point manager (CPM) 连接点管理程序
connection related function (CRF) 连接相关的功能
connection release 连接释放
connection time-out 连接暂停
connectivity transparency 连通（性）透明性
connectless service 无连接的服务
connect-oriented service 面向连接的服务

CONP (connection-oriented network protocol) 面向连接网络协议
consolidated lin layer management (CLLM) 统一链路层管理协议
constant bit rate (CBR) 恒位速率
Consultative Committee of International Radio (CCIR) 国际无线电咨询委员会
Consultative Committee on International Telegraph and Telephone (CCITT) 国际电报电话咨询委员会
consumer digital subscriber line (CDSL) 消费者数字用户线
consumer to business (C to B, C2B) 消费者对企业
consumer to consumer (C to C, C2C) 消费者对消费者
content addressable computing system 内容寻址的计算系统
content addressable method 按内容寻址法
content aggregator 内容聚集商
content cache 内容缓存
content charging gateway (CCG) 内容计费网关
content confidentiality 内容保密性
content conversion 内容转换
content delivery 内容传递
content delivery management 内容发送管理
content delivery network (CDN) 内容传递网,内容发布网
content distribution 内容分发
content distribution network (CDN) 内容分发网络

content filtering 内容过滤
contention 争用，竞争
contention access 争用存取，竞争接入
contention access period (CAP) 争用接入周期
contention-collision cancellation access mode 争用-冲突淘汰访问方式
contention-free access 无争用存取
contention-loser session 冲突失败方会话
contention mode 争用方式
contention network 争用网络
contention PODA (CPODA) 争用的面向优先级按需分配信道
contention ring 争用环
contention ring protocol 争用环网协议
contention-winner session 冲突胜利方会话
content item element 内容项要素
content management (CM) 内容管理
content management server 内容管理服务器
content management system (CMS) 内容管理系统
content page 内容页
content provider 内容提供商
content retrieval 目录检索
content richness 内容丰富性
content scrambling system 内容扰乱保密系统
content specific site 特定内容网站
content type 内容类型
content verification attack 内容验证攻击
continuous ARQ protocol 连续 ARQ 协议，连续自动重发请求协议
continuous phase frequency shift keying (CP-FSK) 连续相位频移键控
continuous phase modulation 连续相位调制
continuous RQ protocol 连续 RQ 协议，连续重发请求协议
continuous wavelet neural network (CWNN) 连续小波神经网络
control and provisioning of wireless access points (CAPWAP) protocol specification 无线接入点控制和配置协议规范
control channels (CCH) 控制信道
control connections 控制连接
control frame 控制帧
control frame packet 控制帧分组
controller area network (CAN) 控制器局域网
controller area network (CAN) bus 控制器局域网总线
control radio network controller (CRNC) 控制无线网络控制器
control signaling 控制信令
control station 控制站
converged network 融合网络
converged network adapter (CNA) 融合网络适配器
convergence sublayer (CS) 会聚子层
convolutional neural network (CNN) 卷积神经网络
conversation 会话
cookie file 点心文件
. coop 非赢利合作机构域名
Cooperation for Open Systems Interconnection Networking in Europe (COSINE)

欧洲开放系统互联网络协会
cooperative intelligence network system 协作智能网络系统
cooperation of Web caching 网络缓存协作
cooperative network 协作网络
cooperative relay network 协作中继网络
coordinative reasoning network 协作推理网络
cooperative research network 协作研究网络
cooperative wireless network 协作无线网络
coordinate multiple points (CoMP) 协作多点
Coordinating Committee for Intercontinental Research Networks (CCIRN) 洲际研究网络协调委员会
copper distributed data interface (CDDI) 铜线分布数据接口
COPS-PR (common open policy service usage for policy provisioning) 公共公共策略服务的策略提供功能
CORBA (common object request broker architecture) 公共对象请求代理体系结构
core based trees (CBT) 有核树
core gateway 核心网关
core network (CN) 核心网
core router 核心路由器
corporate area network (CAN) 企业区域网
corporate blog 企业博客
corporate portals 公司门户站点
corporate website marketing 公司网站营销
corporate website optimization (CWO) 公司网站优化
CoS (class of service) 服务级(别)
COSE (common open software environment) 通用开放软件环境
COSE (common open system environment) 通用开放系统环境
COSINE (Cooperation for Open Systems Interconnection Networking in Europe) 欧洲开放系统互联网络协会
cost per click (CPC) 按点击数付费
cost per lead (CPL) 按引导数付费
cost per sale (CPS) 按销售额付费
counter cipher mode with block chaining message authentication code protocol (CCMP) 计数器模式密码块链消息完整码协议
country code 国家码
country code top level domain names (ccTLD) 国家代码顶级域名
country-or-network identity 国家或网络标识符
courtesy copy (CC) 抄送
CP (connection point) 连接点
CPC (cost per click) 按点击数付费
CP capabilities 控制点能力
CPC (cognitive pilot channel) 感知导频信道
CPCId (common physical channel identifier) 公共物理信道标识
CP-CP sessions 控制点-控制点会话
CPCS (common part convergence sublayer) 公共部件会聚子层

CPCS-SDU (common part convergence sublayer-service data unit) 公共部件会聚子层-服务数据单元
CPE (customer premises equipment) 用户驻地设备
CP-FSK (continuous phase frequency shift keying) 连续相位频移键控
CPL (cost per lead) 按引导数付费
CPM (connection point manager) 连接点管理程序
CPN (calling party number) 调用团体号
CPN (customer premises network) 用户驻地网
CP name 控制点名
CPNI (customer proprietary network information) 用户优先网络信息
CPODA (contention PODA) 争用的面向优先级按需分配信道
CPS (cells per second) 每秒信元数
CPS (commodity promotion solution) 商品推广解决方案
CPS (cost per sale) 按销售额付费
CPS (common part sublayer) 公共部分子层
CPU protect policy (CPP) CPU 保护策略
CR (cognitive radio) 认知无线电
CR (candidate recommendation) 候选推荐标准
Cracker 骇客,闯入者
CRD (call rerouting distribution) 呼叫重选路由分配
credit message 信用报文
CRL (certification revocation list) 证书注销表
CRM (cell rate margin) 信元率边缘
CRM (center resource management) 中心资源管理
CRNC (control radio network controller) 控制无线网络控制器
CRNC communication context CRNC 通信上下文
cross-domain 交叉域
cross-domain keys 跨域密钥
cross-domain network manager session 跨域网络管理器会话
cross-domain resource (CDRSC) 跨域资源
cross-domain resource manager (CDRM) 跨域资源管理器
cross-domain subarea link 跨(地)域子区链路
cross-network 跨网络
cross-network session 跨网络会话
cross office trunk 汇接局中继
cross over cable 交叉电缆
cross over network 分频网络
cross parity checking code 交叉奇偶检验码
cross point array 交换单元阵列
cross post 交叉邮寄
cross site link 交叉点链接
cross site request forgery (CSRF) 跨站请求伪造(攻击)
cross site scripting (XSS) 跨站脚本(攻击)
cross-subarea 跨子区的
cross-subarea link 跨子区链路
crossware 交叉件

crowdfunding 众筹
crowdsourcing 众包
crowdsourcing model 众包模式
CRS（cell relay service） 信元中继服务
crypto currency 密码货币
cryptographic session 密码对话
cryptographic session key 密码对话密钥
CS（convergence sublayer） 会聚子层
C/S（client/server） 客户机/服务器
CSA（Cloud Security Alliance） 云安全联盟
CSAC（Cyber Security Association of China） 中国网络空间安全协会
CSC（circuit switching center） 电路交换中心
CSC（customer service chat） 客户聊天服务
CSCD（Chinese Science Citation Database） 中国科学引文数据库
CSCW（computer-supported collaboration work） 计算机支持的协作工作
CSD（circuit switch data） 电路交换数据
CSDN（circuit switched data network） 电路交换数据网
CSF（client signal fail） 客户端信号失效
CS-GW（circuit switched gateway） 电路交换网关
CSMA（carrier sense multiple access） 载波监听多路访问
CSMA/CA（carrier sense multiple access/collision avoid） 载波监听多路访问/冲突避免

CSMA/CD（carrier sense multiple access/collision detection） 载波监听多路访问/冲突检测
CSMA/CD access method CSMA/CD 访问方法
CSMA/CD network CSMA/CD 网络
CS-MUX（circuit switching multiplexer） 电路交换多路复用器
CSN（circuit switched network） 电路交换网络
CSRF（cross site request forgery） 跨站请求伪造
CSRZ（carrier suppressed return zero） 载波抑制归零码
CSS（cascading style sheets） 层叠样式表单
CSTNET（China Science and Technology Network） 中国科技网
CSV（circuit-switched voice） 电路交换语音
CSV/CSD（alternate circuit-switched voice/circuit-switched data） 候选电路交换话音与电路交换数据
CTCId（common transport channel identifier） 公共传输信道标识
CTD（cell transfer delay） 信元传输延迟
CTI（computer telephony image） 三网合一
CTI（computer telecommunication integration） 计算机电信集成
CTL（certificate trust list） 证书信用［托管］表
C to B（consumer to business） 消费者对企业

CTP (connectionless transport protocol) 无连接传输协议

CTP (command terminal protocol) 命令终端协议

CTP (centralized token passing) 集中式令牌传送

CTR (click through ratio) 点击通过率

CU (See You) 再见

CUD (call user data) 呼叫用户数据

CUL (see you later) 下次再会

culture of network consumption 网络消费文化

current cell rate (CCR) 当前信元速率

customer access 用户接入

customer bridged network 用户桥接网络

customer controlled reconfiguration/rerouting (CCR) 用户控制重配置/重路由选择

customer device circuits 用户设备线路

customer edge (CE) 用户边缘设备

customer installation (CI) 用户设备装置

customer interaction network 客户交互网络

customer premises equipment (CPE) 用户驻地设备

customer premises network (CPN) 用户驻地网

customer premises network service 用户驻地网业务

customer proprietary network information (CPNI) 用户优先网络信息

customer service chat (CSC) 客户聊天服务

customer station equipment 客户站设备

customized application for mobile enhanced logic (CAMEL) 移动网增强逻辑的定制应用

cut-and-paste attack 剪贴攻击

cut-through frame switching 穿透式帧交换

cut-through packet switching 穿透式分组交换

cut-through switching 穿透式交换

CVDT (cell variation delay tolerance) 信元时延变动的允许范围

CVE (common vulnerabilities and Exposures) 通用漏洞披露(组织)

CVE (collaborative virtual environments) 协作虚拟环境

CVM (collaborative virtual manufacturing) 协作虚拟制造

CWDM (coarse wavelength division multiplexing) 稀疏波分复用

CWIS (campus wide information system) 园区信息服务系统

CWNN (continuous wavelet neural network) 连续小波神经网络

CWNP (certified wireless networking professional) 无线网络专家认证

CWO (corporate website optimization) 公司网站优化

cXML (commerce XML) 商用扩展置标语言

cyber- 赛博

cyber acquaintance 网友

cyberaddict 网迷

cyber age 网络时代

cyber attack   网络攻击
cyberbullying   网络欺凌
cyberburger joint   网络快餐店
cybercafe   网络咖啡屋,网吧
cybercash   网络现金,电子现金
cyber chat   网络聊天
cybercitizen   网络公民,网民
cybercity   网络城市,虚拟城市
cybercollege   网上大学
cybercop   电脑警察
cybercrime   电脑犯罪,网络犯罪
cybercult   网迷,网虫
cyberculture   赛博文化,网络文化
cyber democracy   网络民主
cyberdepot   网络商店
cyber exposure   网络曝光,网曝
cybereconomy   网络经济,互联网经济
cyberfair   网络集市,网络展览会
cyberfiction   赛博虚构,网络小说
cyberforensics   网络取证,数字取证
cyber forger   网络造假者,网上伪造者
cyber handle   网络匿名
cyber heroine   网络美女
cyber intruder   网络入侵者,黑客
cyberized   网络化
cyber jihad   网络圣战
cyberjock   网迷
cyber journalism   网络新闻
cyberlanguage   网络语言
cyberlover   网络情人
cyber mall   网上超市
cyber marketing   网络营销
cybernaut   网客,网迷
cybernated   电脑的,网络化
cybernation   自动控制的,网络化

cybernetist   控制论专家
cybernews   网络新闻
cyber moral violence   网络道德暴力
cyber newsletter   网上简讯
cyberpet   网络宠物,虚拟宠物
cyberphobia   电脑恐惧症
cyberpicketing   网络警戒
cyber politics   网络政治
cyberprise   网络企业
cyberpub   网上酒店
cyberpunk   电脑朋客,电脑高手
cyber Romeo   网上罗密欧
cybersalon   网络沙龙
cybersatirist   网上讽刺家
Cyber Security Association of China (CSAC)   中国网络空间安全协会
cybersex   网络色情
cybershop   网上商店,网店
cyber shopper   网上导购员
cybersoap   网络肥皂剧
cyberspace   赛博空间,网络空间
cyberspeak   网络用语
cyber squatting   域名抢注
cybersquatter   域名抢注者
cybersquatting   域名抢注
cyber station   网站
cyberstalker   网络骚扰者
cybersurfing   网络漫游,网上冲浪
cyber terrorism   网络恐怖主义
cyber violence   网络暴力
CyberWar   网络战争
cyberwidow   网络寡妇
cyber wonk   网络迷
cyber word   网络用语
cyberworld   网络世界,虚拟世界

**cyberyakker** 网上垃圾制造者
**cyborg** 赛伯格,电子人
**cybrarian (cyberspace librarian)** 网络图书馆员,信息搜索者
**cypher punk** 密码朋克
**C2B (consumer to business)** 消费者对企业
**C2C (consumer to consumer)** 顾客对顾客
**c2cwitkey** 点对点威客
**c2cwitkey marketing** 点对点威客营销
**C3 revolution (Computation、Control、Communication)** 3C革命
**C4 (command, control, computer, and communication)** 指挥、控制、计算机和通信

# D

DaaS (data as a service) 数据即服务
DAD (duplicate address detection) 地址重复检测
daily activited users (DAU) 日活跃用户
daily kernel activited users (DKAU) 日核心活跃用户数
daily login users (DLU) 日登录用户数
daily new users (DNU) 日新增用户数
DAMA (demand assignment multiple access) 按需分配多址
DAN (DNS acceleration network) 域名系统加速网络
dark Web 暗 Web,模糊 Web
data above voice (DAV) 话上数据
data above voice transmission 话上数据传输
data access arrangement (DAA) 数据存取阵列
data acquisition information 数据采集信息
data as a service (DaaS) 数据即服务
database concurrent operation 数据库并发操作
database gateway 数据库网关
database server 数据库服务器
data broadcast (DB) 数据广播
data carrier detect (DCD) 数据载波检测
data channel (D channel) 数据通道,D信道
data channel controller 数据通道控制器
data circuit switching system 数据电路转接系统
data circuit terminating equipment (DCE) 数据电路终端设备
data circuit terminating equipment (DCE) waiting 数据电路终端设备等待
data circuit transparency 数据电路透明性
data coding 数据编码
data collision 数据冲突
data communication (DC) 数据通信
data communication adapter 数据通信适配器
data communication channel (DCC) 数据通信通道
data communication equipment (DCE) 数据通信设备
data communication function (DCF) 数据通信功能
data communication interface 数据通信接口
data communication network (DCN)

数据通信网

data communication protocol 数据通信协议

data communication service 数据通信业务

data communication system 数据通信系统

data country code (DCC) 数据国家码

data cross connect (DCC) 数字交叉连接

data diffusion 数据扩散

data diffusion leaks 数据扩散泄密

data-driven attack 数据驱动攻击

data encapsulation 数据封装

data encryption algorithm (DEA) 数据加密算法

data encryption standards (DES) 数据加密标准

data exchange interface (DXI) 数据交换接口

data flow analyse 数据流分析

data flow control (DFC) 数据流控制

data flow control (DFC) layer 数据流控制层

data flow control protocol 数据流控制协议

data flow model 数据流模型

data frame packet 数据帧包

datagram 数据报

datagram congestion control protocol (DCCP) 数据拥塞控制协议

datagram delivery protocol (DDP) 数据报传送协议

datagram network 数据报网络

datagram nondelivery indication 数据报不能传递指示

datagram service 数据报服务(程序)

datagram service signal 数据报服务信号

datagram socket (DS) 数据报套接字

datagram transport layer security (DTLS) 数据报传输层安全(协议)

data highway 数据高速通道

data host node 数据宿主节点

data information frame 数据信息帧

data information packet 数据信息包

data island 数据孤岛

data island integration 数据孤岛集成

data island phenomenon 数据孤岛现象

data link connection identifier (DLCI) 数据链路连接标识符

data link control (DLC) 数据链路控制

data link control layer 数据链路控制层

data link control protocol (DLCP) 数据链路控制协议

data link control type 数据链路控制类型

data link laycr (DLL) 数据链路层

data link layer protocol 数据链路层协议

data link layer packet 数据链路层信息包

data link layer protocol specification 数据链路层协议规范

data link layer service 数据链路层服务

data link layer topology discovery 数据

链路层拓扑发现
data link switching (DLSW) 数据链路交换
data mining 数据挖掘
data mining agent 数据挖掘代理
data mining extensions (DMX) 数据挖掘扩展插件
data mining model 数据挖掘模型
data mining technology 数据挖掘技术
data network 数据网络
data network identification code (DNIC) 数据网络标识码
data over voice (DOV) 话上数据,语音加载数据
data packet 数据分组[包]
data packet dropout 数据分组丢失
data packet format 数据分组格式
data packet interception 数据分组截获
data packet switching network 数据分组交换网
data packet transmitting 数据分组传输
data processing node 数据处理节点
data redundancy reduction (DRR) 冗余数据简化
data reference frame 数据基准帧
data service unit (DSU) 数据服务单元
data signaling rate transparency 数据信号速率的透明性
data stream mining 数据流挖掘
data switching equipment (DSE) 数据交换设备
data tampering 数据篡改
data terminal equipment (DTE) 数据终端设备
data transparency 数据透明性
data under voice (DUV) 话下数据
DAU (daily activited users) 活跃用户
DAV (data above voice) 话上数据
DB (data broadcast) 数据广播
DBA (dynamic bandwidth allocation) 动态带宽分配
DBL (domain block list) 域名谱遏制列表
DBRMP (dynamic broadcast ring multicast routing protocol) 动态广播环组播路由协议
DC (domain controller) 域控制器
DCE (data circuit terminating equipment) 数据电路终端设备
DCE (data communication equipment) 数据通信设备
DCE (distributed computing environment) 分布计算环境(标准)
DCF (data communication function) 数据通信功能
DCF (distributed coordination function) 分布式协调功能
DCF interframe space (DIFS) DCF帧间间隔
D channel (data channel) 数据通道,D信道
DCH (dedicated channel) 专用信道
DCL (digital channel link) 数字信道连接
DCN (data communication network) 数据通信网
DCNA (data communication network architecture) 数据通信网络体系

结构
DCS (distributed computer system) 分布式计算机系统
DCS (digital cross-connect system) 数字交叉连接系统
DD (delay distortion) 延迟失真
DDCMP (digital data communication message protocol) 数字数据通信报文协议
DDN (digital data network) 数字数据网
DDN (defense data network) 国防数据网
DDN (dotted decimal notation) 点分十进制计数法
DDNS (dynamic domain name system) 动态域名系统
DDNS (dynamic domain name server) 动态域名服务器
DDoS (distributed denial of service) 分布式拒绝服务
DDoS mitigation service (DMS) 抵御DDoS攻击服务
DDP (datagram delivery protocol) 数据报传送协议
DDR (dial-on-demand routing) 按需拨号路由技术
DDS (digital domain system) 数字域名系统
DDSA (digital data service adapter) 数字数据服务适配器
DEA (data encryption algorithm) 数据加密算法
dead-letter box 死信信箱
dead-letter file 死信文件

dead link 死链接
deceptive adware 欺骗性广告软件
dedicated access 专用访问
dedicated channel (DCH) 专用信道
dedicated control channel (DCCH) 专用控制信道
dedicated line 专用线路
dedicated packet data group (DPG) 专用分组数据群
dedicated physical channel (DPCH) 专用物理信道
deep convolution neural network 深度卷积神经网络
deep packet inspection 深层数据包检测
deep stateful packet inspection 深度数据包检测
deep Web 深度Web,深网
deep Web crawling 深网抓取
deep Web database 深网数据库
deep web entry 深度网入口
deep Web information 深网信息
default activity network 默认活动网络
default gateway 默认网关
default gateway address 默认网关地址
default home page 默认主页
default mode network 默认模式网络
default network 默认网络
default network message queue 默认网络消息队列
default network output queue 默认网络输出队列
default node representation 默认节点表示
default receive windows 预设接收窗口

default route  默认路由
default route configuration  默认路由配置
default SSCP list  默认 SSCP 表
defense data network (DDN)  国防数据网
defense information infrastructure common operating environment (DIICOE)  国防信息基础设施公共操作环境
deferring procedure  延迟过程
deficit round robin (DRR)  差额循环调度
definite response (DR)  确认响应
delay ACK  延迟确认
delay distortion (DD)  延迟失真
delayed response mode  延迟应答方式
delay modulation (DM)  延迟调制
delay of service  服务延迟
delay queue  延迟队列
delay tolerant network (DTN)  时延容忍网络，容迟网络
Delta routing  δ 路由选择
demand allocation channel  按需分配信道
demand assignment multiple access (DAMA)  按需分配多址
demand priority  请求优先级
demand priority access (DPA)  需求优先访问
demand priority access LAN  需求优先存取局域网
demand priority access method (DPAM)  需求优先存取方法
demand priority MAC protocol  需求优先介质访问控制协议
demand priority protocol (DPP)  需求优先协议
demand priority working group  需求优先工作组
democratically synchronized network  等权同步网络
democratic network  等权网
democratic network mutually synchronized network  等权网互同步网
demultiplex (DEMUX)  多路解调
demultiplexing  多路分解
demultiplexing detection  多路分解检测
DEMUX (demultiplex)  多路解调
DEN (directory enabled network)  目录允许网络
denial of service (DoS)  拒绝服务（攻击）
dense wavelength division multiplexing (DWDM)  密集波分复用
departmental server  部门级服务器
DES (data encryption standard)  数据加密标准
DES (destination end station)  目标终结站
designated bridge  指定网桥
designated router (DR)  指定路由器
despotically synchronized network  主时钟控制同步网
DEST (destination address)  目的地址
destination address (DEST)  目的地址
destination address field (DAF)  目的地址字段
destination address routing  目的地址路由选择

destination cache table  目的地缓存表
destination dialing  目的地直拨方式
destination end station (DES)  目标终结站
destination logical unit (DLU)  目标逻辑单元
destination MAC address (DA)  目标介质访问控制地址
destination service access point (DSAP)  目标服务访问指针
destination subarea field (DSAF)  目标子区域
detection of the wrong data package  数据包丢失检测机制
deterministic LAN  确定性局域网
DFWMAC (distrubuted foundation wireless medium access control)  分布式基础无线网介质访问控制
DHCP (dynamic host configuration protocol)  动态主机配置协议
DHCP domain  DHCP 域
DHCP relay agent  DHCP 中继代理
DHTML (dynamic hyper text markup language)  动态超文本标记语言
diagnostic and monitoring protocol (DMP)  诊断与监控协议
dialog control  对话控制
dialog unit  对话单元
dial-on-demand routing (DDR)  按需拨号路由技术
dial peer  拨号点
dial peer hunting  拨号点搜寻
dial-up access  拨号上网
dial-up account  拨号账户
dial-up control  电话拨号控制

dial-up modem  拨号调制解调器
Diameter base protocol (DBP)  Diameter 基础协议
Diameter cryptographic message syntax (DiameterCMS)  Diameter 密码消息语法协议
Diameter extensible authentication protocol (DiameterEAP)  Diameter 可扩展鉴别协议
Diameter mobile IP (DiameterMIP)  Diameter 移动 IP 协议
Diameter network access service (DiameterNAS)  Diameter 网络接入服务协议
Diameter protocol  Diameter 协议
differential Manchester encoding  差分曼彻斯特编码
differential quadrature phase shift keying (DQPSK)  差分四相相移键控
differentiated service (DiffServ)  区分服务
differentiated service code point (DSCP)  差分服务代码点
differentiated service model  区分服务模型
differentiated service network  区分服务网
DiffServ (differentiated service)  区分服务
DIFS (DCF interframe space)  DCF 帧间间隔
Digg  掘客
Digg website  掘客网站
digerati  计算机或网络行家
digest authentication  摘要式验证

digital asset 数字资产
digital asset management (DAM) 数字资产管理
digital audio broadcast network 数字音频广播网络
digital broadcast video network 数字广播视频网络
digital campus network 数字校园网络
digital cash 数字现金
digital cellular mobile communications network 数字蜂窝移动通信网
digital certificate 数字证书
digital certificate identification 数字证书认证
digital channel link (DCL) 数字信道连接
digital city 数字城市
digital communication 数字通信
digital cross-connect equipment (DXC equipment) 数字交叉连接器
digital cross-connect system (DCS) 数字交叉连接系统
digital data communication message protocol (DDCMP) 数字数据通信报文协议
digital data network (DDN) 数字数据网
digital data service adapter (DDSA) 数字数据服务适配器
digital data switching 数字数据交换
digital defense 数字防御
digital divide 数字鸿沟
digital domain 数字域名
digital domain system (DDS) 数字域名系统

digital earth 数字地球
digital economy 数字经济
digital fingerprinting 数字指纹
digital fingerprinting protocol 数字指纹协议
digital footprint 数字脚印
digital inheritance 数字遗产
digital interactive media 数字互联媒介
digital jihad 数字圣战
digital library 数字图书馆
digital loop carrier (DLC) 数字环路载波
digital marketing 数字营销
digital multiplex hierarchy 数字复用体系
digital narrowcast network 数字窄播网络
digital narrowcast system 数字窄播系统
digital network 数字网络
digital network architecture (DNA) 数字网络体系结构
digital network economy 数字化网络经济
digital network environment 数字网络环境
digital object unique identifier (DOI) 数字对象唯一识别符
digital pulse wireless 数字脉冲无线电
digital right management (DRM) 数字版权管理
digital satellite channel 数字卫星信道
digital service unit (DSU) 数字服务单元

digital service unit-radio (DSU-R) 无线数字服务单元
digital set top box converter 数字机顶盒转换器
digital signature 数字签名
digital signature algorithm (DSA) 数字签名算法
digital signature certificate 数字签名证书
digital signature mechanism 数字签名机制
digital signature standard (DSS) 数字签名标准
digital simultaneous voice and data (DSVD) 数字语音数据同传
digital subscriber line (DSL) 数字用户线路
digital subscriber line (DSL) broadband access technology 数字用户线宽带接入技术
digital subscriber network 数字用户网络
digital telephone network 数字电话网
digital time-stamp service (DTS) 数字时间戳服务
digital TV library 数字电视图书馆
digital video 数字视频
digital video network 数字视频网络
digital wallets 数字钱包
digital watermark 数字水印
digital wireless local loop 数字无线本地环路
digital wrappers 数字包装器,数字包裹
digiteer 计算机能手

digit head 计算机迷
DIIK (damned if I known) 我真的不知道
direct address mapping 直接地址映射
direct attached storage (DAS) 直连存储
directed broadcast 定向广播
directed broadcast address 定向广播地址
direct network connection 直接网络连接
directory access protocol (DAP) 目录访问协议
directory enabled network (DEN) 目录允许网络
directory replication 目录复制
directory routing 目录式路由选择
directory service (DS) 目录服务
directory service markup language (DSML) 目录服务标记语言
directory system protocol (DSP) 目录系统协议
directory user agent (DUA) 目录用户代理
discrete multi-tone (DMT) 离散多音（调制）
disjoint network 分离网
disjoint path network 不相交路径网络,独立路径网络
DISL (dynamic interior switching link) 动态交换链路内协议
disruption tolerant network (DTN) 中断容忍网络
distance education 远程教育
distance learning 远程学习

distance vector multicast routing protocol (DVMRP) 距离向量组播路由协议

distance vector routing 距离向量路由选择

distance vector routing protocol 距离向量路由协议

distributed adaptive routing 分布式自适应路由选择

distributed adaptive routing algorithm 分布式自适应路由选择算法

distributed agent approach 分散代理法

distributed architecture (DA) 分布式体系结构

distributed asynchronous fusion 分布式异步融合

distributed asynchronous Web services component 分布式异步Web服务组件

distributed computer system (DCS) 分布式计算机系统

distributed computer system network 分布式计算机系统网络

distributed computing 分布式计算

distributed computing environment (DCE) 分布计算环境(标准)

distributed computing network 分布式计算网络

distributed concurrency control 分布式并行控制

distributed connectionist network 分布连接网络

distributed control network 分布式控制网络

distributed coordination function (DCF) 分布式协调功能

distributed data acquisition network 分布式数据采集网络

distributed data mining 分布式数据挖掘

distributed data processing network 分布式数据处理网络

distributed data stream mining 分布式数据流挖掘

distributed decision support system (DDSS) 分布式决策支持系统

distributed denial of service (DDoS) 分布式拒绝服务

distributed deterministic routing algorithm 分布式确定路由算法

distributed digital control technology 分布式数字控制技术

distributed directory database 分布式目录数据库

distributed file system (DFS) 分布式文件系统

distributed file system server 分布式文件系统服务器

distrubuted foundation wireless medium access control (DFWMAC) 分布式基础无线网介质访问控制

distributed gateway 分布式网关

distributed hash index 分布式散列索引

distributed hierarchical routing 分布式层次路由

distributed interactive simulation (DIS) 分布交互仿真

distributed LAN switch 分布式局域网交换机

distributed logon security 分布式登录安全性

distributed loop computer network 分布式环形计算机网络

distributed management environment (DME) 分布式管理环境

distributed management of resource information 资源信息分布式管理

distributed media access control 分布式介质访问控制

distributed message-switching system 分布式报文交换系统

distributed minicomputer network 分布式小型计算机网络

distributed multicast routing 分布式组播路由

distributed naming service 分布式命名服务

distributed network 分布式网络

distributed network agent (DNA) 分布式网络代理

distributed network computing 分布式网络计算

distributed network parallel computing 分布式网络并行计算

distributed network system (DNS) 分布式网络系统

distributed object computing (DOC) 分布对象计算

distributed object network management 分布式对象网络管理

distributed office support system (DISOSS) 分布式办公支持系统

distributed operating system 分布式操作系统

distributed optical fiber sensor 分布式光纤传感器

distributed optical fiber sensor network 分布式光纤传感器网络

distributed orthogonal routing 分布式正交路由

distributed parallel file system 分布式并行文件系统

distributed processing 分布(式)处理(方式)

distributed processing network architecture 分布式处理网络体系

distributed processing system 分布式处理系统

distributed reflection denial of servie (DRDoS) 分布式反射拒绝服务(攻击)

distributed resilient routing 分布式弹性路由

distributed robot control system 分布式机器人控制系统

distributed routing 分布式路由选择

distributed sensor network 分布式传感器网络

distributed service network (DSN) 分布式业务网络

distributed spanning tree (DST) 分布式生成树

distributed supervisor control system 分布式监控系统

distributed system architecture (DSA) 分布式系统体系结构

distributed system network 分布式系统网络

distributed testing control system 分布

式测控系统
distributed transaction 分布式事务
distributed transaction coordinator protocol 分布式事务协调协议
distributed transaction management 分布事务管理
distributed transaction processing services 分布式事务处理服务
distribution parameter network 分布式参数网络
distribution queue dual bus (DQDB) 分布式队列双总线
distribution service network 分布式服务网络
distribution tree 分布树
dividing network 分频网络,选频网络
diving water 潜水
DIX Ethernet (Digital Intel Xerox Ethernet) DIX以太网
DKAU (daily kernel activited users) 日核心活跃用户数
DKIM (domain keys identified mail) 域名密钥识别邮件
DL (downlink) 下行
DLC (digital loop carrier) 数字环路载波
DLC (data link control) 数据链路控制
DLCI (data link connection identifier) 数据链路连接标识符
DLCP (data link control procedure) 数据链路控制协议
DLL (data link layer) 数据链路层
DLSW (data link switching) 数据链路交换

DLU (daily login users) 日登录用户数
DME (distributed management environment) 分布式管理环境
DMS (DDoS mitigation service) 抵御DDoS攻击服务
DNA (digital network architecture) 数字网络体系结构
DNA (distributed network agent) 分布式网络代理
DNAT (dynamic network address translation) 动态网络地址转换
DNIC (data network identification code) 数据网络标识码
DNS (domain name system) 域名系统
DNS (distributed network system) 分布式网络系统
DNS (domain name server) 域名服务器
DNS acceleration network (DAN) 域名系统加速网络
DNS amplification attack DNS放大攻击
DNSSEC (domain name system security) 域名系统安全协议
DNS spoofing DNS欺诈
DNU (daily new users) 日新增用户数
document object model (DOM) 文档对象模型
document type definition (DTD) 文档类型定义
DOI (digital object unique identifier) 数字对象唯一识别符
DOM (document object model) 文档对象模型
domain block list (DBL) 域名遏制

列表
domain controller (DC)　域控制器
domainer　域名人，玉米
domain for sale　域名出售
domain keys identified mail (DKIM)　域名密钥识别邮件
domain name (DN)　域名
domain name activity　域名活跃度
domain name dispute　域名争议
domain name dispute policy　域名争议规则
domain name dispute resolution　域名争议解决
domain name dispute resolution mechanism　域名争议解决机制
domain name hijacking　域名劫持
domain name inverse mapping　域名反向映像
domain name inverse query　域名反向查询
domain name pointer query　域名指针查询
domain name registrar　域名注册商
domain name registration　域名注册
domain name registration queries　域名注册查询
domain name resolution　域名转换，域名解析
domain name server (DNS)　域名服务器
domain name system (DNS)　域名系统
domain name system poisoning　域名系统投毒
domain name system security (DNSSEC)　域名系统安全协议

domain search　领域搜索
domain server　域服务器
dominant operator　主导运营商
doorway pages　门户网面，门页
DoS (denial of service)　拒绝服务（攻击）
dot address　带点地址
dotted decimal　点十进制数
dotted decimal notation (DDN)　点分十进制计数法
double opt-in E-mail marketing　双向确认邮件营销
double probabilistic neural network　双概率神经网络
double token access mode　双令牌存取方式
DOV (data over voice)　话上数据，语音加载数据
downlink (DL)　下行
downlink bandwidth　下行带宽
downlink shared channel (DSCH)　下行共享信道
download　下载
downloading acceleration　下载加速
download manager　下载管理
download software　下载软件
downstream neighbor　下游邻居
DPA (demand priority access)　需求优先访问
DPAM (demand priority access method)　需求优先存取方法
DPCH (dedicated physical channel)　专用物理信道
DPP (demand priority protocol)　需求优先协议

DQPSK(differential quadrature phase shift keying) 差分四相相移键控
DR(dynamic reconfiguration) 动态重新配置
DR(definite response) 确认响应
DR(designated router) 指定路由器
DRDoS(distributed reflection denial of servie) 分布式反射拒绝服务攻击
dredge 爬网
drift radio network controller(DRNC) 漂移无线网络控制器
DRNC(drift radio network controller) 漂移无线网络控制器
DRR(data redundancy reduction) 冗余数据简化
DS(datagram socket) 数据报套接字
DS(directory service) 目录服务
DS(dunce smiley) 笨伯
DSA(directory system agent) 目录系统代理
DSA(distributed system architecture) 分布式系统体系结构
DSAP(destination service access point) 目标服务访问指针
DSCH(downlink shared channel) 下行共享信道
DSCP(differentiated service code point) 差分服务代码点
DSL(digital subscriber line) 数字用户线路
DSL access multiplexer(DSLAM) 数字用户线路接入复用器
DSLAM(digital subscriber line access multiplexer) 数字用户线路接入复用器

DSML(directory service markup language) 目录服务标记语言
DSN(distributed service network) 分布式业务网络
DSP(directory system protocol) 目录系统协议
DSR(dynamic source routing) 动态源路由选择
DSU-R(digital service unit-radio) 无线数字服务单元
DSVD(digital simultaneous voice and data) 数字语音数据同传
DS3 physical layer convergence protocol(DS3 PLCP) DS3物理层会聚协议
DS3 PLCP(DS3 physical layer convergence protocol) DS3物理层会聚协议
DTD(document type definition) 文档类型定义
DTE(data terminal equipment) 数据终端设备
DTLS(datagram transport layer security) 数据报传输层安全
DTM(dynamic synchronous transfer mode) 动态同步传送模式
DTN(delay tolerant network) 时延容忍网络
DTN(disruption tolerant network) 中断容忍网络
DTP(dynamic trunking protocol) 动态中继协议
DTS(digital time-stamp service) 数字时间戳服务
DUA(directory user agent) 目录用户代理

dual-attached station (DAS) 双向连接站
dual attachment concentrator (DAC) 双连接集中器
dual attachment connection 双连连接
dual attachment port 双连端口
dual-cable broadband LAN 双缆宽带局域网
dual homed host gateway 双宿主机网关
dual homing 双连接入，双宿主
dual hubbed 双枢纽
dual IP layer 双重 IP 层
dual leaky bucker algorithm 双漏桶算法
dual node interconnection 双节点互连
dual node interconnection protection 双节点互连保护
dual node pair 对偶结点对
dual protocol stack 双协议栈
dual stack (DS) 双协议栈，双栈
dual stack DNS 双栈域名系统
dual stack mobile IP 双栈移动 IP
dual-stack operation 双栈操作
dual-stack router 双栈路由器
dual stack transition mechanism (DSTM) 双栈过渡机制
dual velocity leaky bucker algorithm 双速漏桶算法
dummy backward channel 反向假信道
duplicate address detection (DAD) 地址重复检测
DUV (data under voice) 话下数据
DVMRP (distance vector multicast routing protocol) 距离向量组播路由协议
DWDM (dense wavelength division multiplexing) 密集波分复用
dynamic adaption routing 动态自适应寻径
dynamic address 动态地址
dynamic address allocation protocol 动态地址分配协议
dynamic address mapping 动态地址映射
dynamic address relocation 动态地址重定位
dynamic address resolving 动态地址解析
dynamic address translation 动态地址翻译
dynamic address translator 动态地址转换器
dynamically established data link 动态建立数据链路
dynamic alternate routing 动态备选路由
dynamic adaptive routing 动态自适应路由
dynamic bandwidth 动态带宽
dynamic bandwidth controller 动态带宽控制器
dynamic congestion control algorithm 动态拥塞控制算法
dynamic convergence network 动态收敛网络
dynamic network bandwidth adjusting 动态网络带宽调节
dynamic bandwidth allocation (DBA) 动态带宽分配

dynamic bandwidth allocation algorithm 动态带宽分配算法
dynamic broadcast ring multicast routing protocol（DBRMP） 动态广播环组播路由协议
dynamic channel allocation（DCA） 动态信道分配
dynamic channel assignment routing 动态信道分配路由
dynamic control function 动态控制功能
dynamic convergence network 动态收敛网络
dynamic data website 动态数据网站
dynamic dispatch virtual table 动态配送虚拟表
dynamic domain name resolution 动态域名解析
dynamic domain name server（DDNS） 动态域名服务器
dynamic domain name service 动态域名服务
dynamic domain name system（DDNS） 动态域名系统
dynamic forensics of network intrusion 网络入侵动态取证
dynamic fuzzy neural network（OFNN） 动态模糊神经网络
dynamic fuzzy semantic network 动态模糊语义网
dynamic hierarchical network routing 动态分级网络路由选择
dynamic host configuration protocol（DHCP） 动态主机配置协议
dynamic hyper text markup language（DHTML） 动态超文本标记语言
dynamic intelligent agent 动态智能代理软件
dynamic interior switching link（DISL） 动态交换链路内协议
Dynamic IP packet filtering 动态IP包过滤
dynamic IP routing lookup 动态IP路由查找
dynamic mobile host configuration protocol（DHCP） 动态移动主机配置协议
dynamic mobile IP key update 动态移动IP密钥更新
dynamic monitoring network 动态监测网络
dynamic network 动态网络
dynamic network address translation（DNAT） 动态网络地址转换
dynamic network architecture 动态网络体系结构
dynamic network database 动态网络数据库
dynamic network database query system 动态网络数据库查询系统
dynamic neuron network 动态神经元网络
dynamic node ID assignment 动态节点ID赋值
dynamic node planning 动态节点规划
dynamic optimum route 动态最优路由
dynamic optimal route guidance 动态最优路由诱导
dynamic overall bandwidth allocation 动态总体带宽分配

dynamic packet filter 动态包过滤器
dynamic packet filtering firewall 动态包过滤防火墙
dynamic packet transport (DPT) 动态分组传输
dynamic path update 动态路径更新
dynamic ranking 动态排名
dynamic reconfiguration 动态重新配置
dynamic recurrent fuzzy neural network 动态递归模糊神经网络
dynamic recurrent network 动态递归网络
dynamic recurrent wavelet neural network 动态递归小波神经网络
dynamic root node 动态根节点
dynamic route switching 动态路由变换
dynamic routing 动态路由选择
dynamic routing and wavelength assignment 动态路由和波长分配
dynamic routing protocol 动态路由协议
dynamic semantic network 动态语义网
dynamic sensor network 动态传感器网络
dynamic source routing (DSR) 动态源路由选择
dynamic source routing protocol 动态源路由协议
dynamic source routing strategy 动态源路由策略
dynamic synchronous transfer mode (DTM) 动态同步传送模式
dynamic threshold query 动态门限查询
dynamic translation 动态转换
dynamic virtual trap network 动态虚拟陷阱网络
dynamic virtual public network 动态虚拟公共网络
dynamic vision sensor network 动态视觉传感器网络
dynamic Web 动态 Web
dynamic Web GIS 动态的 Web 地理信息系统
dynamic Web page 动态网页
dynamic website 动态网站
dynamic wireless network 动态无线网
dynamic wireless network formation protocol 动态无线组网协议
dynamic wireless sensor network 动态无线传感器网络

# E

EAP (extensible authentication protocol) 可扩展认证协议
early token release 提前释放令牌
EB (electronic business) 电子商务
EBGP (external border gateway protocol) 外部边界网关协议
E-business 电子商务
E-business website 电子商务网站
EC (electronic commerce) 电子商务
E-Cash (electronic cash) 电子现金
E-channel E 信道
E-Check (electronic check) 电子支票
echo frame (ECF) 回应帧
echo reply 回波应答
echo request 回波请求
ECM (enterprise content management) 企业内容管理
ECMR (equal cost multipath routing) 等价多路径路由
ECN (explicit congestion notification) 显式拥塞指示
ECN (emergency communics network) 应急通信网络
E-commerce 电子商务
E-coupons 电子优惠券
ECP (enterprise customer portal) 企业客户门户
ECS (enterprise communication server) 企业通信服务器
EC2 (elastic compute cloud) 弹性计算云
EDGE (enhanced data rate for GSM evolution) 增强型数据速率 GSM 演进技术
edge access router 边缘接入路由器
edge computing 边缘计算
edge device 边缘设备
edge media router 边缘媒体路由器
edge network 边缘网络
edge of network (EON) 网络边际
edge router 边缘路由器
edge service router 边缘业务路由器
edge weighted network 边缘赋权网络
EDI (electronic data interchange) 电子数据交换
EDM (electronic direct mail advertising) 电子直邮广告
EDP (extreme discovery protocol) 极进发现协议
EDTR (effective data transfer rate) 有效数据传送率
.edu 教育机构域名
education entertainment (edutainment) 寓教于乐的软件
education software 教学软件
edutainment (education entertainment)

寓教于乐的软件

EEA（Enterprise Ethereum Alliance）企业以太坊联盟

EFCI（explicit forward congestion indication） 显式前向冲突指示

effective data transfer rate（EDTR） 有效数据传送率

effect of information island 信息孤岛效应

efficient TCP（ETCP） 智能TCP

EFM（Ethernet for the first mile） 以太网最前一英里

EFS（error free seconds） 无错时间

EFT（electronic funds transfer） 电子汇兑

E-Government（electronic government）电子政务

EGP（exterior gateway protocol） 外部网关协议

EGP features 外部网关协议特点

EGP measure distances 外部网关协议计量距离

E-Group（electronic group） 电子小组

EIA（Electronic Industries Association）（美国）电子工业协会

EIA/CCITT（Electronics Industries Association/International Telegraph and Telephone Consultative Committee）（美国）电子工业协会/国际电报电话咨询委员会

EIA/TIA（Electronics Industries Association/Telecommunication Industries Association）（美国）电子工业协会/电信工业协会

EIGRP（enhanced interior gateway routing protocol） 增强型内部网关路由协议

EII（enterprise information infrastructure）企业信息基础设施

EIP（enterprise information portal） 企业信息门户

EIP（enterprise application portal） 企业应用门户

e-jihad 电子圣战

ELAN（emulated local area network）仿真局域网

elastic compute cloud（EC2） 弹性计算云

elastic compute cloud service 弹性计算云服务

elasticity buffering 弹性缓冲

electric control optical packet switching 电控光分组交换

elastic cloud 弹性云端

elastic cloud compute 弹性云计算

elastic cloud host 弹性云主机

elastic cloud server 弹性云服务器

electronic advertisement 电子广告

electronic bank 电子银行

electronic bank transaction for home 家庭电子银行业务

electronic business（EB） 电子商务

electronic business platform 电子商务平台

electronic business solution 电子商务解决方案

electronic cash（E-Cash） 电子现金

electronic check（E-Check） 电子支票

electronic commerce（EC） 电子商务

electronic commerce model 电子商务

模式
electronic currency 电子货币
electronic data interchange (EDI) 电子数据交换
electronic direct mail advertising (EDM) 电子直邮广告
electronic game 电子游戏
electronic government (E-Government) 电子政务
electronic group (E-Group) 电子小组
Electronic Industries Association (EIA) (美国)电子工业协会
electronic journal 电子期刊
electronic library 电子图书馆
electronic magazine (E-zine) 电子杂志
electronic mail (E-mail) 电子邮件
electronic mail protocol 电子邮件协议
electronic mail system (EMS) 电子邮件系统
electronic marketplace 电子市场
electronic marketplace server 电子市场服务器
electronic money 电子货币
electronic newspapers 电子报纸
electronic press 电子报刊
electronic purse 电子钱包
electronic society E[电子]社会
electronic terrorism 电子恐怖主义
electronic virtual marketplace 电子虚拟市场
electronic wallet (E-wallet) 电子钱包
electro-optic modulator (EOM) 电光调制器
E-magazine 电子期刊
E-mail (electronic mail) 电子邮件
E-mail address 电子邮件地址
E-mail advertising 电子邮件广告
E-mail affair 网上爱情
E-mail alias 电子邮件别名
E-mail authentication 电子邮件认证
E-mail box 电子邮件信箱
E-mail direct marketing (EDM) 电子邮件营销
E-mail filter 电子邮件过滤器
E-mail netiquette 电子邮件网络礼仪
E-mail server 电子邮件服务器
E-mail standards 电子邮件标准
E-mail system model 电子邮件系统模型
E-mail types of messages 电子邮件报文类型
E-mail virus 电子邮件病毒
embedded dynamic Web 嵌入式动态 Web
embedded network 嵌入式网络
embedded network communication 嵌入式网络通信
embedded network computing 嵌入式网络计算
embedded network instrument 嵌入式网络仪器
embedded network module 嵌入式网络模块
embedded network security 嵌入式网络安全
embedded network single-chip computer 嵌入式网络单片机
embedded network technology 嵌入式网络技术
embedded network video server 嵌入式

网络视频服务器
embedded Web　嵌入式网站
embedded Web platform　嵌入式 Web 平台
embedded Web service　嵌入式 Web 服务
embedded Web server　嵌入式 Web 服务器
embedded video Web　嵌入式视频 Web
embedded heterogeneous network　嵌入式异构网络
emergency call network　应急呼叫网络
emergency communics network (ECN)　应急通信网络
emergency monitoring network　应急监测网络
emergency network　应急网络
e-monitor　电子监视器
emotag　情感标志
Emoticon (emotion icon)　表情符号
emotion icon (Emoticon)　表情符号
emotion symbol　情感符号
EMP (Ethernet management port)　以太网管理端口
EMS (electronic mail system)　电子邮件系统
EMS (enhanced message service)　增强型短信服务
emulated local area network (ELAN)　仿真局域网
encapsulating security payload (ESP)　封装安全载荷
encapsulation bridging　封装网桥
end of transmission block (ETB)　信息块传送终止

end system-to-end system (ES-to-ES) communication　末端系统之间通信
end system to intermediate system (ES-IS)　末端系统到中间系统
end-to-end encryption　端到端加密
end-to-end network　端到端网络, 对等网络
end-to-end protocol　端到端协议
enhanced control area network　增强型控制局域网
enhanced data rate for GSM evolution (EDGE)　增强型数据速率 GSM 演进技术
enhanced interior gateway routing protocol (EIGRP)　增强型内部网关路由协议
enhanced logical link control (ELLC)　增强型逻辑链路控制
enhanced message service (EMS)　增强型短信服务
enhanced multi-level precedence and pre-emption service (EMLPP)　增强型多级优先和占先业务
enhanced network　增强网络
enhanced network survivability performance　增强网络可生存性性能
enhanced partial network coding　增强型局部网络编码
enhanced service provider (ESP)　增强业务提供商
enhanced telecom operations map (eTOM)　增强型电信运营图
E-NNI (exterior-network to network interface)　外部网络-网络接口
entangled quantum effect　纠缠量子

效应
extended border node  扩展边界节点
enterprise application integration（EAI）企业应用集成
enterprise application portal（EIP）企业应用门户
enterprise bank  企业银行
enterprise blog  企业博客
enterprise blog marketing  企业博客营销
enterprise business website  企业商务网站
enterprise communication server（ECS）企业通信服务器
enterprise computer network  企业网络；企业计算机网络
enterprise content management（ECM）企业内容管理
enterprise cooperative network  企业协作网络
enterprise customer portal（ECP）企业客户门户
enterprise E-business  企业电子商务
Enterprise Ethereum Alliance（EEA）企业以太坊联盟
enterprise external network  企业外部网络
enterprise extranet  企业外联网
enterprise information infrastructure（EII）企业信息基础设施
enterprise information portal（EIP）企业信息门户
enterprise internal network  企业内部网络
enterprise mailbox  企业邮箱
enterprise marketing network  企业营销网络
enterprise mobile application platform  企业移动应用平台
enterprise mobile portal  企业移动门户
enterprise network  企业网络
enterprise network directory service  企业网络目录服务
enterprise network operating system  企业网络操作系统
enterprise portal  企业门户
enterprise search platforms（ESP）企业搜索平台
enterprise server  企业级服务器
enterprise service bus（ESB）企业服务总线
enterprise system connection（ESCON）企业系统连接
enterprise virtual switching network（EVISN）企业虚拟智能交换网
enterprise website  企业网站
enterprise website hosting  企业网站寄存
enterprise-wide network  企业范围网络
entry border node  入口边界节点
environment specific inter ORB protocol（ESIOP）特定环境对象请求代理间协议
EON（edge of network）网络边际
EP（error protocol）错误协议
EPON（Ethernet passive optical network）以太网无源光网络
EPR（electronic public relation）网络

公关
equal cost multipath routing (ECMR) 等价多路径路由, 等价路由
ER (explicit route)　显式路由
ERPS (Ethernet ring protection switching) 以太网环保护交换
error free seconds (EFS)　无错时间
error lock　出错封锁
error protocol (EP)　错误协议
ERTE (explicit route table entry)　显式路由表项
ESB (enterprise service bus)　企业服务总线
ESF (extended superframe format)　扩展超帧格式
ESF (extended service frame)　扩展服务帧
ESIOP (environment specific inter ORB protocol)　特定环境对象请求代理间协议
ES-IS (end system to intermediate system)　末端系统到中间系统
ESMTP (extended simple mail transfer protocol)　扩展简单邮件传送协议
ESP (encapsulating security payload)　封装安全载荷
ESP (enterprise search platforms)　企业搜索平台
ESP (enhanced service provider)　增强业务提供商
ESS (extended service set)　扩展服务群
established connection　建立连接
ETCP (efficient TCP)　智能 TCP
E-text　电子文本
ETH (Ethereum)　以太坊, 以太币

Ethereum (ETH)　以太坊, 以太币
Ethereum Foundation　以太坊基金会
Ethereum virtual machine (EVM)　以太坊虚拟机
Ethernet　以太网
Ethernet access switch　以太网接入交换机
Ethernet card　以太网卡
Ethernet communication interface　以太网通信接口
Ethernet controller　以太网控制器
Ethernet converter　以太网转换器
Ethernet for the first mile (EFM)　以太网最前一英里
Ethernet hosts　以太网主机
Ethernet interface　以太网接口
Ethernet interface node　以太网接口节点
Ethernet line monitor　以太网线路监控器
Ethernet local module　以太网本地模块
Ethernet management port (EMP)　以太网管理端口
Ethernet meltdown　以太网熔化
Ethernet passive optical network (Ethernet-PON, EPON)　以太网无源光网络
Ethernet-PON (Ethernet passive optical network)　以太网无源光网络
Ethernet port　以太网端口
Ethernet port groups　以太网端口组
Ethernet ring protection switching (ERPS)　以太网环保护交换
Ethernet serial interface　以太网络串

行接口
Ethernet service switch　以太网服务交换机
Ethernet switch　以太网交换机
Ethernet transceiver　以太网传输器
Ethernet transmission cable　以太网传输电缆
ethical hacker　道德黑客
ethical hacking　道德黑客攻击
eTOM（enhanced telecom operations map）　增强型电信运营图
EVISN（enterprise virtual switching network）　企业虚拟智能交换网
EVM（Ethereum virtual machine）　以太坊虚拟机
E-wallet　电子钱包
exception request（EXR）　异常请求
exception response（EX）　异常应答
exchange data link　交换数据链路
exchange ethernet network　交换式以太网
exchange link alliance　交换链接联盟
exchange network service　交换网络服务
exchange office　交换局
exchange partner　换客
exclusive caching　专属缓存
exclusive piconet　独享微微网
exhaustive misrouting backtracking　穷举错误路由回溯
exhaustive profitable algorithm　穷举有利算法
exhaustive route search　穷举路由搜索
explicit congestion notification（ECN）　显式拥塞指示
explicit connection　显式连接
explicit forward congestion indication（EFCI）　显式前向拥塞指示
explicit route（ER）　显式路由
explicit route compression　显式路由压缩
explicit route length　显式路由长度
explicit route table　显式路由表
exploits block list（XBL）　拓展遏制列表
exposed node　暴露节点
exposed node problem　暴露节点问题
exposed terminal　暴露终端
express transfer protocol（XTP）　快速传输协议
extended active virtual network management　扩展的主动虚拟网络管理
extended border node　扩展边界节点
extended interframe space（EIFS）　延长帧间间隔
extended network　扩展网络
extended network addressing　扩展网络地址
extended packet reservation multiple access　扩充分组预留多址访问
extended queuing network　扩展的排队网络
extended reala-time polling service（ertPS）　扩展实时轮询服务
extended response byte　扩充响应字节
extended role network models　角色网络扩展模型
extended route　扩充路由
extended semantic network　扩展语义

网络
extended service frame (ESF) 扩展服务帧
extended service set (ESS) 扩展服务群
extended simple mail transfer protocol (ESMTP) 扩展简单邮件传送协议
extended subarea addressing 扩展的子区域寻址
extended superframe format (ESF) 扩展超帧格式
extended terminal access contorller access control system (XTACACS) 扩展终端访问控制器访问控制系统
extensible authentication protocol (EAP) 可扩展认证协议
extensible link language (XLL) 可扩展链接语言
extensible markup language (XML) 可扩展标记语言
extensible messaging and presence protocol (XMPP) 可扩展消息和呈现协议
extensible provisioning protocol (EPP) 可扩展供应协议
extensible stylesheet language (XSL) 可扩展样式语言
extension station 扩充站
extensive domain name 泛域名
extensive domain name resolution 泛域名解析
exterior gateway 外部网关

exterior gateway protocol (EGP) 外部网关协议
exterior link 外部链路
exterior-network to network interface (E-NNI) 外部网络-网络接口
exterior packet switch 外部包交换
exterior reachable address 外部可到达地址
exterior route 外部路由
external border gateway protocol (EBGP) 外部边界网关协议
external link 外部链接
external link anchor text 外部链接锚文本
external link state table 外部链路状态表
external modulation 外调制
external reference synchronization 外部参照同步
external route summarization 外部路由汇总
external routing protocol 外部路由协议
extranet 外联网
extranet login 外联网登录
extro network node interface protocol 外部网络节点接口协议
Extreme discovery protocol (EDP) 极进发现协议
E-zine (electronic magazine) 电子杂志

# F

FACH（forward access channel） 前向接入信道
FACCH （ fast association control channel） 快速随路控制信道
Facebook 脸谱网
failure domain 失效区域
fair arbitration 公平仲裁
fair non-repudiation protocol （FNP） 公平非抵赖协议
fair queuing（FQ） 公平排队
FAN（fiber access network） 光纤接入网
fan-out unit 扇出单元
fans 粉丝
FAR（fixed alternate routing） 固定备选路由选择
FAS（frame alignment signal） 帧同步信号
fast association control channel （FACCH） 快速随路控制信道
fast circuit switching（FCS） 快速电路交换
fast connect circuit switching 快速连接电路交换
fast crossbar switch network 高速交叉开关网络
fast Ethernet 快速以太网
fast Ethernet switch link 快速以太网交换机间链路
fast forward 快速转发
fast frequency-shift keying（FFSK） 快速频移键控
fast industrial Ethernet 快速工业以太网
fast LAN switch 快速局域网交换机
fast local area network 快速局域网
fast low-latency access with seamless hand-off OFDM（Flash OFDM） 无线传输低延迟访问正交频分复用
fast packet switch（FPS） 快速分组交换
fast mobile IP（FMIP） 快速移动 IP
fast network fault location 快速网络故障定位
fast switch Ethernet 快速交换以太网
fast turn around（FTA） 快速转换（控制位）
fat client 胖客户机
fat client model 胖客户机模型
fat pipe 胖管道
fat server model 胖服务器模型
fat server 胖服务器
fat tree 胖树
fat tree network 胖树网络
fat tree topology 胖树拓扑
fault management（FM） 故障管理

fault tolerant computing 容错计算
fault tolerant controller 容错控制器
fault tolerant design 容错设计
fault tolerant local area network 容错局域网
fault tolerant mechanism 容错机制
fault tolerant network 容错网络
fault tolerant network computing 容错网络计算
fault tolerant operation 容错运行,容错操作
fault tolerant routing 容错路由选择
fault tolerant routing algorithm 容错路由算法
fault tolerant software 容错软件
fault tolerant Web service 容错 Web 服务
favorites 收藏夹
fax mail 传真邮件
FBR 基于流量路由选择
FBWA (fixed broadband wireless access) 固定宽带无线接入
FC (fiber channel) 光纤信道
FCAPS (fault, configuration, accounting, performance, security) 网络的五大管理功能
FCC (Federal Communications Commission) (美国)联邦通信委员会
FCIP (fiber channel over IP) 在 IP 上的光纤信道
FCS (frame check sequence) 帧检验序列
FCS (fast circuit switching) 快速电路交换
FCSI (fiber channel system initiative) 光纤通道系统倡议组织
FCSN (fiber channel storagenetwork) 光纤通道存储网络
FDD (frequency division duplex) 频分双工
FDAM (flexible data access model) 柔性数据访问模型
FDDI (fiber distributed data interface) 光纤分布数据接口
FDDI (fiber distributed data interchange) 光纤分布数据交换
FDDI/CDDI 光纤分布数据接口/铜线分布数据接口
FDDI MAC sublayer protocols FDDI 介质访问控制子层协议
FDDI network constraint 光纤分布数据接口网络限制
FDDI-Ⅱ Ⅱ型光纤分布数据接口
FDHP (full duplex handshaking protocol) 全双工握手协议
FDM (frequency division multiplexing) 频分复用
FDMA (frequency division multiple access) 频分多址
FDN (flow distribution network) 流量分发网络
FDSE (full-duplex switched Ethernet) 全双工交换以太网
FE (for example) 举例
feasible flow 可行流
feasible circulation flow 可行循环流
feasible fuzzy flow 可行模糊流
FEBE (far end block error) 远端块错误
FEBE (front end/back end) 前端/

后端

FEC (forwarding equivalence class) 转发等价类

FEC (forward-error correction) 正向纠错

FECN (forward explicit congestion notification) 前向显式拥塞通告

Federal Communications Commission (FCC) （美国）联邦通信委员会

Federal Information Processing Standards (FIPS) （美国）联邦信息处理标准

Federal Networking Council (FNC) （美国）联邦网络委员会

Federal Research Internet Coordinating Committee (FRICC) （美国）联邦因特网研究协作委员会

feedback neural network 反馈神经元网络

FEFO (first ended, first out) 先结束先出

FEP (firewall enhancement protocol) 防火墙增强协议

FEP (front-end processor) 前端处理机

FGP (fuzzy grid partition) 模糊网格划分

FH-CDMA (frequency hopping code division multiple access) 跳频码分多址

FHRP (first hop redundancy protocol) 首跳冗余性协议

FHSS (frequency hopping spread spectrum) 跳频展频

fiber access network (FAN) 光纤接入网

fiber bandwidth 光纤带宽

fiber channel (FC) 光纤信道

fiber-channel arbitrated loop (FC-AL) 光纤信道仲裁环

fiber channel over IP (FCIP) 在 IP 上的光纤信道

fiber channel storagenetwork (FCSN) 光纤通道存储网络

Fiber Channel System Initiative (FCSI) 光纤通道系统倡议组织

fiber connector (FICON) 光纤连接器

fiber cross connector (FXC) 光纤交叉连接器

fiber distributed data interface (FDDI) 光纤分布数据接口

fiber HSLN 光纤高速局部网

fiber in the loop (FITL) 光纤用户环路

fiber optic cable concentrator 光缆集中器

fiber optic channel 光纤通道

fiber optic communication (FOC) 光纤通信

fiber optic communication network 光纤通信网

fiber optic data bus 光纤数据总线

fiber optic data transfer network 光纤数据传输网

fiber optic demultiplexer (active) 光纤多路分解器(有源)

fiber optic demultiplexer (passive) 光纤多路分解器(无源)

fiber optic interrepeater link (FOIRL) 中继器间光纤链路

fiber optic link (FOL) 光纤链路
fiber optic loop multiplexer 光纤环路复用器
fiber optic mixer 光纤混频器
fiber optic modem 光纤调制解调器
fiber optic multiplexer (active) 光纤复用器(有源)
fiber optic multiplexer (passive) 光纤复用器(无源)
fiber optic network 光纤网络
fiber optic transmission system (FOTS) 光纤传输系统
fiber to the building (FTTB) 光纤到大楼
fiber to the curb (FTTC) 光纤到路边
fiber to the home (FTTH) 光纤到家(庭)
fiber to the LAN (FTTLAN) 光纤到局域网
fiber to the node (FTTN) 光纤到节点
fiber to the premises (FTTP) 光纤到驻地
fifteen ones convention 15个一规则
file locking 文件锁定
file server 文件服务器
file sharing protocol 文件共享协议
file signature verification (FSV) 文件签名验证
file transfer 文件传送
file transfer access and management (FTAM) 文件传送存取与管理
file transfer protocol (FTP) 文件传输协议
file transfer protocol server 文件传输协议服务器

file transmission path 文件传输路径
filtered symmetric differential phase-shift keying (FSDPSK) 过滤对称差分相移键控
financial service markup language (FSML) 金融服务标记语言
firewall 防火墙
firewall configuration 防火墙设置
firewall detection 防火墙探测
firewall principle 防火墙原理
firewall safety 防火墙安全
first-ended first-out (FEFO) 先结束先送出
first hop redundancy protocol (FHRP) 首跳冗余性协议
first hop router 首跳路由器
FITL (fiber in the loop) 光纤用户环路
fixed alternate routing (FAR) 固定备选路由选择
fixed broadband wireless access (FBWA) 固定宽带无线接入
fixed fiber optic connector 固定光纤连接器
fixed length subnet mask (FLSM) 固定长度子网掩码
fixed mobile convergence (FMC) 固网移动融合
fixed routing (FR) 固定路由选择
fixed wavelength optical add drop multiplexer (FW-OADM) 固定波长光分插复用器
fixed wireless access (FWA) 固定无线接入
flag of frame 帧标志

flame bait  网络激怒诱饵，网络论战
flamer  网上令人讨厌的人
flame war  不良信息战，网络论战
flash update  闪速更新
flash video (FLV)  FLV 格式，流媒体格式
flat name space  平面名字空间
flat naming  平面命名机制
flat network  平坦网络
flexible access control policy  灵活访问控制策略
flexible access system (FAS)  灵活接入系统
flexible access termination  灵活接入终端
flexible contact network  柔性接点网
flexible data access  柔性数据访问
flexible data access model (FDAM)  柔性数据访问模型
flexible information transport capability  柔性信息传输能力
flexible network  柔性网络
flexible neural network  柔性神经网络
flexible point (FP)  灵活点
flexible service  柔性服务
floating ad  漂浮广告
floating static route  浮动静态路由
flood  滥发
flooding attack  泛洪攻击
flooding routing  泛洪路由选择
flood search routing  扩散式搜索路由选择
flow-based routing (FBR)  基于流量路由选择
flow control  流量控制
flow control digit  流控制数
flow control methods  流控方法
flow control packets  数据流控制信息包
flow control procedure  流量控制规程
flow distribution network (FDN)  流量分发网络
flow label  流标记
flowtable  流表
FLSM (fixed length subnet mask)  固定长度子网掩码
flush protocol  帧流协议
FMD (function management data)  功能管理数据
FMH (function management header)  功能管理标题
FMIP (fast mobile IP)  快速移动 IP
FNC (Federal Networking Council)  （美国）联邦网络委员会
FOC (fiber optic communication)  光纤通信
FOCH (forward channel)  正向信道
fog computing  雾计算
follow-up posting  继承式消息投递
foreign address  外部地址
formal node name  正式节点名字
forward access channel (FACH)  前向接入信道
forward and backward reasoning  正反向推理
forward channel (FOCH)  正向信道
forward caching  正向缓存
forward chained reasoning  正向链推理
forward direction  前向

forward echo 正向回波
forward-error correction (FEC) 正向纠错
forward explicit congestion notification (FECN) 前向显式拥塞通告
forwarding 转发
forwarding equivalence class (FEC) 转发等价类
forward LAN channel 前向 LAN 通道
forward link 正向链接
forward reasoning 正向推理
forward reasoning network 正向推理网络
forward regular reasoning 正向规则推理
fourth generation mobile communication technology (4G) 第四代移动通信技术
FP (flexible point) 灵活点
FP (frame protocol) 帧协议
FPLMTS (future public land mobile telecommunication system) 未来公用陆地移动通信系统
fps (frames per second) 帧每秒
FPS (fast packet switch) 快速分组交换
FR (frame relay) 帧中继
FR (fixed routing) 固定路由选择
fractional T1 部分 T1
fractional T1 service (FT1) 部分 T1 电路业务
fractional T3 service (FT3) 部分 T3 电路业务
FRAD (frame relay assembler/disassembler) 帧中继汇集/分散器

fragmentation threshold 分割界限
fragment-free switching 自由分段交换
frame 帧
frame alignment pattern 帧定位格式
frame alignment signal (FAS) 帧同步信号
frame chaining 帧链接
frame check sequence (FCS) 帧检验序列
frame control 帧控制
frame control field 帧控制字段
frame-control window 帧控制窗口
framed interface 帧界面
frame duration 帧持续时间
frame errors 帧差错
frame filtering 帧过滤
frame format 帧格式
frame frequency 帧频
frame level 帧级
frame maintenance mode 帧维持方式
frame page 帧页面
frame protocol (FP) 帧协议
frame reject 帧拒绝
frame relay (FR) 帧中继
frame relay access device (FRAD) 帧中继接入设备
frame relay access support (FRAS) 帧中继接入支持
frame relay assembler/disassembler (FRAD) 帧中继汇集/分散器
frame relay frame 帧中继帧
frame relay frame handler (FRFH) 帧中继帧处理器
frame relay network 帧中继网

frame relay service (FRS) 帧中继服务
frame relay terminal equipment (FRTE) 帧中继终端设备
frame slip 帧滑移
frame source file 框架源文件
frame start delimiter 帧起始指示
frame structure 帧结构
frame switcher 帧交换机
frame switching technology 帧交换技术
frame synchronization 帧同步
frame synchronization sequence 帧同步序列
frame synchronizer 帧同步器
frame type (FT) 帧类型
frame window 帧窗口
framing 组帧
framing bit 成帧位,帧划分位
FRAS (frame relay access support) 帧中继接入支持
free buffer enquiry 空闲缓冲区查询包
free domain 免费域名
free homepage space 免费主页空间
FreeMail 免费邮箱
free network address element 空闲网络地址元素
freenet 自由网
free proxy server 免费代理服务器
free routing 自由路由选择
free space optical communication (FSOC) 自由空间光通信
free Web homepage space 免费主页空间
free website hosting 免费网站托管

frequency and optical wavelength converter (FWC) 频率和光波长转换器
Frequency Assignment Authority (FAA) 频率分配机构
frequency averaging 频率均衡
frequency band allocation 频段分配
frequency band transmission 频带传输
frequency carrier 载波频率,载频
frequency change signaling (FCS) 换频信令
frequency-derived channel 频分信道
frequency discriminator 鉴频器
frequency diversity (FD) 频率分集
frequency division 分频
frequency division duplex (FDD) 频分双工
frequency division multiple access (FDMA) 频分多址
frequency division multiplexing (FDM) 频分复用
frequency division multiplexing (FDM) standards 频分复用标准
frequency division switching 频分交换
frequency hopping code division multiple access (FH-CDMA) 跳频码分多址
frequency hopping spread spectrum (FHSS) 跳频展频
frequency section selection repeater 频率部分选择中继器
frequency-shift keyed modulation (FSKM) 频移键控调制
frequency-shift keying (FSK) 频移键控
frequently asked questions (FAQ) 常见

问题

FRFH (frame relay frame handler) 帧中继帧处理器

FR forum frame relay standards 帧中继论坛帧中继标准

FRICC (Federal Research Internet Coordinating Committee) (美国)联邦因特网研究协作委员会

front-end network processor (FNP) 前端网络处理机

front-end processor (FEP) 前端处理机

front-end software 前端软件

front interchange node 前置交换结点

FRS (frame relay service) 帧中继服务

FRTE (frame relay terminal equipment) 帧中继终端设备

FSAN (full service access network) 全业务接入网

FSDPSK (filtered symmetric differential phase-shift keying) 过滤对称差分相移键控

FSK (frequency shift keying) 频移键控

FSKM (frequency-shift keyed modulation) 频移键控调制

FSN (full service network) 全业务网络

FSOC (free space optical communication) 自由空间光通信

FT (frame type) 帧类型

FTF (face to face) 面对面

FTP (file transfer protocol) 文件传输协议

FTP client ftp 客户程序

FTP commands FTP 命令

FTP server FTP 服务器

FTP site FTP 站点

FTTB (fiber to the building) 光纤到大楼

FTTC (fiber to the curb) 光纤到路边

FTTH (fiber to the home) 光纤到家

FTTN (fiber to the node) 光纤到节点

FTTO (fiber to the office) 光纤到办公室

FTTP (fiber to the premises) 光纤到驻地

FT1 (fractional T1 service) 部分 T1 电路业务

FT3 (fractional T3 service) 部分 T3 电路业务

full baud bipolar coding (FBBC) 全波特双极性编码

full-duplex Ethernet 全双工以太网

full duplex handshaking protocol (FDHP) 全双工握手协议

full-duplex line 全双工线路

full-duplex switched Ethernet (FDSE) 全双工交换以太网

full fibre network 全光纤网络

full gateway 全网关

full interconnected network 全互连网络

full IP core network 全 IP 核心网

full mesh 全网型

full mobility 全移动性

full IP network 全 IP 网络

full IP network architecture 全 IP 网络结构

full node name 节点全名

full series network 全面服务网络
full service access network（FSAN） 全业务接入网
full service intranet 全服务内联网
full service network（FSN） 全业务网络
fully automated network design 全自动网络设计
fully connected cubic network 全互连立方体网络
fully connected network 全连通网络
fully qualified domain name（FQDN） 全称域名
functional signaling 功能信令
functional signaling link 功能信令链接
function management data（FMD） 功能管理数据
function management header（FMH） 功能管理标题
function management layer 功能管理层
function management profile 功能管理轮廓文件
function protocol layer 功能协议层
fusion of information island 信息孤岛融合
future public land mobile telecommunication system（FPLMTS） 未来公用陆地移动通信系统
fuzzy adaptive robust fault tolerant control 模糊自适应鲁棒［健壮］容错控制
fuzzy artificial neural network 模糊人工神经网络

fuzzy clustering neural network 模糊聚类神经网络
fuzzy competitive neural network 模糊竞争神经网络
fuzzy control network 模糊控制网络
fuzzy evaluating neural network 模糊评价神经网络
fuzzy grid partition（FGP） 模糊网格划分
fuzzy inference network 模糊推理网络
fuzzy inference neural network 模糊推理神经网络
fuzzy leaky bucket algorithm 模糊漏桶算法
fuzzy modular neural network 模糊模块化神经网络
fuzzy network 模糊网络
fuzzy neural network control 模糊神经网络控制
fuzzy neural network diagnosis model 模糊神经网络诊断模型
fuzzy quantum neural computational network 模糊量子神经计算网络
fuzzy quantum neural network 模糊量子神经网络
fuzzy reasoning network 模糊推理网络
fuzzy reasoning neural network 模糊推理神经网络
Fuzzy wavelet neural network 模糊小波神经网络
FWA（fixed wireless access） 固定无线接入
FWC（frequency and optical wavelength converter） 频率和光波长转换器

**FW-OADM**（fixed wavelength optical add drop multiplexer） 固定波长光分插复用器
**FXC**（fiber cross connector） 光纤交叉连接器
**FYA**（for your amusement） 供娱乐
**FYI**（for your information） 供参考

# G

GAA (generic authentication architecture) 通用认证架构
game client 游戏客户端
game currency 游戏币
game operator 游戏运营商
GAP (gatekeeper) 网守,网闸
garbage 垃圾数据
gatekeeper (GAP,GK) 网守
gateway 网关
gateway access protocol 网关接入协议
gateway access request 网关访问请求
gateway access service 网关接入服务
gateway-capable host 可作网关的宿主机
gateway control function 网关控制功能
gateway discovery protocol (GDP) 网关发现协议
gateway facilities 网关设备
gateway-gateway protocol (GGP) 网关-网关协议
gateway GPRS supporting node (GGSN) 网关通用分组无线业务支持节点
gateway host 网关主机
gateway interface 网关接口
gateway load balancing 网关负载均衡
gateway load balancing protocol (GLBP) 网关负载均衡协议
gateway mobile services switching center 网关移动业务交换中心
gateway mobile switching center (GMSC) 网关移动交换中心
gateway network control program (gateway NCP) 网关网络控制程序
gateway network element 网关网络元件
gateway node 网关节点
gateway nodes model 网关节点模型
gateway node monitor 网关节点监控
gateway requirements 网关需求
gateway serving node 网关服务节点
gateway SSCP 网关系统服务控制点
gateway support node 网关支持节点
gateway switching center 网关交换中心
gateway to gateway protocol (GGP) 网关间协议
gateway VTAM 网关虚拟远程通信访问法
gateway with network station 带网络工作站的网关
Gaussian filtered minimum shift keying (GMSK) 高斯滤波最小频移键控
Gaussian frequency-shift keying (GFSK) 高斯频移键控
Gaussian minimum shift keying (GMSK)

高斯最小频移键控

**GB (gigabyte)** 千兆字节,吉字节

**Gbps (gigabits per second)** 每秒千兆位

**GBR (guaranteed bit rate)** 保证比特率

**Gb/s (gigabits per second)** 每秒千兆位

**GCRA (generic cell rate algorithm)** 通用信元速率算法

**GDCN (generalized dynamic constraints network)** 广义动态约束网络

**GDP (gateway discovery protocol)** 网关发现协议

**GEC (Gigabit Ethernet Consortium)** 千兆位以太网联合会

**GEMS (government electronic marketplace service)** 政府电子市场服务

**geek** 极客

**gender of the network** 网络社会性别

**general data stream (GDS)** 通用数据流

**general inter-ORB protocol (GIOP)** 通用对象请求代理间协议

**generalized activity network** 广义活动网络

**generalized clustering network** 广义聚类网络

**generalized collaboration network** 广义合作网络

**generalized dynamic constraints network (GDCN)** 广义动态约束网络

**generalized dynamic fuzzy neural network** 广义动态模糊神经网络

**generalized path information unit trace (GPT)** 广义路径信息单元轨迹

**generalized network** 广义网络,一般网络

**generalized network simulator** 通用网络模拟器

**generalized transition network** 通用转换网络,广义状态转移网络

**general multiprotocol label switching (GMPLS)** 通用多协议标记交换

**general net theory** 广义网论,通用网络理论

**general network management protocol (GNMP)** 通用网络管理协议

**general packet radio service (GPRS)** 通用分组无线业务

**general pervasive computing** 广义普适计算

**general radio packet service** 通用无线分组业务

**general search engine (GSE)** 通用搜索引擎

**generalized sparse grid** 广义稀疏网格

**general switch management protocol (GSMP)** 通用交换管理协议

**general topology subnetwork** 普通拓扑子网络

**generation gap** 代沟

**generic authentication architecture (GAA)** 通用认证架构

**generic cell rate algorithm (GCRA)** 通用信元速率算法

**generic connection admission control (GCAC)** 通用连接许可控制

**generic controller description** 通用控制器描述

**generic flow control (GFC)** 通用流控制

**generic network information model (GNIM)** 通用网络信息模型

generic routing encapsulation (GRE) 通用路由封装

geographic information system (GIS) 地理信息系统

GEPON (gigabit Ethernet passive optical network) 千兆位以太网无源光网络

GFC (generic flow control) 通用流控制

GFSK (Gaussian frequency-shift keying) 高斯频移键控

GGP (gateway to gateway protocol) 网关间协议

GGSN (gateway GPRS supporting node) 网关通用分组无线业务支持节点

GHz (gigahertz) 千兆赫

giga (G) 千兆,吉

gigabit Ethernet (GE) 千兆位以太网

Gigabit Ethernet Alliance 千兆位以太网联盟

Gigabit Ethernet Consortium (GEC) 千兆位以太网联合会

gigabit Ethernet port 千兆以太网端口

gigabit Ethernet technology 千兆位以太网技术

gigabit fibre optic network 千兆位光纤网络

gigabit Ethernet passive optical network (GEPON) 千兆位以太网无源光网络

gigabit LAN 千兆位局域网

gigabit independent interface (GMII) 千兆位介质无关接口

gigabit network intrusion prevention system (GNIPS) 千兆位网络入侵预防系统

gigabit passive optical network (GPON) 千兆位无源光网

gigabits per second (Gbps) 每秒千兆位

gigabyte (GB) 千兆字节,吉字节

gigabyte system network (GSN) 千兆字节系统网络

gigahertz (GHz) 千兆赫,吉赫

GIG (global information grid) 全球信息网格

GII (global information infrastructure) 全球信息基础设施

GIMM (graded index multimode) 渐变折射率多模(光纤)

GIOP (general inter-ORB protocol) 通用对象请求代理间协议

GIS (geographic information system) 地理信息系统

give a reward 打赏

GK (gatekeeper) 网守

GLBP (gateway load balancing protocol) 网关负载均衡协议

global address 全局地址

global area network 全局地区网络

global digital divide 全球数字鸿沟

global eyes 全球眼

global information grid (GIG) 全球信息网格

global information infrastructure (GII) 全球信息基础设施

globally unique identifier (GUID) 全局唯一标识符

globally unique unicast address 全球唯一单播地址

global mirror 全球镜像

global naming schemes (GNS) 全局命名方案

global naming service 全局命名服务

global network 全球网络

global network addressing domain  全球网络编址域

global network delivery nodel (GNDM)  全球网络交付模式

global network navigator (GNN)  全球网络导航器

global network operation center (GNOC)  全球网络运行中心

global online freedom act (GOFA)  全球网络自由法案

global service logic (GSL)  全局业务逻辑

global system for mobile communication (GSM)  全球移动通信系统

global tracking network  全球跟踪网

global unicast address  全球单播地址

global virtual network service  全球虚拟网业务

GMII (gigabit media independent interface)  千兆位介质无关接口

GMPLS (general multiprotocol label switching)  通用多协议标记交换

GMSC (gateway mobile switching center)  网关移动交换中心

GMSK (Gaussian minimum shift keying)  高斯最小频移键控

GMSK (Gaussian filtered minimum shift keying)  高斯滤波最小频移键控

GNDM (global network delivery nodel)  全球网络交付模式

GNIM (generic network information model)  通用网络信息模型

GNIPS (gigabit network intrusion prevention system)  千兆位网络入侵预防系统

GNMP (general network management protocol)  通用网络管理协议

GNN (global network navigator)  全球网络导航器

GNOC (global network operation center)  全球网络运行中心

GNS (global naming schemes)  全局命名方案

go-back-n ARQ  回退 n 帧的自动重发请求

GOFA (global online freedom act)  全球网络自由法案

gold bridge engineering  "金桥"工程

gold card engineering  "金卡"工程

gold gate engineering  "金关"工程

good 9 (good night)  晚安,再见

Gopher  地鼠,Gopher 检索工具

Gopher client  Gopher 客户

Gopher protocol  Gopher 协议

GOS (grade of service)  服务等级

GOSIP (government open systems interconnection profile)  (美国)政府开放系统互联纲要

.gov  政府机构域名

government affairs client  政务客户端

government affairs micro blog  政务微博

government affairs WeChat  政务微信

government electronic marketplace service (GEMS)  政府电子市场服务

government micro blog  政务微博

government open systems interconnection profile (GOSIP)  (美国)政府开放系统互联纲要

government private network  政务专网

government to business (G2B, G to B)  政府与企业

government to citizen (G2C, G to C)  政府

与公众

government to employee (G2E, G to E) 政府与雇员

government to government (G2G, G to G) 政府与政府

GPON (gigabit passive optical network) 千兆位无源光网

GPON encapsulation method (GEM) GPON 封装模式

GPRS (general packet radio service) 通用分组无线业务

GPRS mobility management and session management GPRS 移动性管理与会话管理

GPRS supporting node (GSN) 通用分组无线业务支持节点

GPRS tunnel protocol (GTP) 通用分组无线业务隧道协议

grade of service (GOS) 服务等级

grand orphan 孤儿进程

grant rights mask 代理权限屏蔽

graphnet 图形传输网

gray field 灰场

GRE (generic routing encapsulation) 通用路由封装

green field 绿场

green network game 绿色网游

gremlin 小捣蛋鬼

grid computing 网格计算

grid node 网格节点

grid node service 网格节点服务

grid service mining (GSM) 网格服务挖掘

grid storage 网格存储

grid system 网格系统

group addressing 成组编址

group address message 组地址信息

group chat 群聊

group delay distortion 群时延失真

group format identifier (GFI) 群格式标识符

group link 群链路

group management protocol 管理协议

group modulation 群调制

group of logical unit 逻辑单元组

Group Special Mobile (GSM) 移动通信特别小组

group synchronization 群同步

groupware 群件

GSE (general search engine) 通用搜索引擎

GSL (global service logic) 全局业务逻辑

GSM (global system for mobile communication) 全球移动通信系统

GSM (group special mobile) 移动通信特别小组

GSMP (general switch management protocol) 通用交换管理协议

GSN (GPRS supporting node) 通用分组无线业务支持节点

GSM (grid service mining) 网格服务挖掘

GSN (gigabyte system network) 千兆字节系统网络

G to B (government to business) 政府与企业

G to C (government to citizen) 政府与公众

G to E (government to employee) 政府与雇员

G to G (government to government) 政

府与政府

**GTP（GPRS tunnel protocol）** 通用分组无线业务隧道协议

**GTP（general data transfer platform）** 通用数据传输平台

**guaranteed bit rate（GBR）** 保证比特率

**guard time** 保护时间

**guest station** 客站

**GUID（globally unique identifier）** 全局唯一标识符

**guidance of public opinion online** 网络舆论引导

**guidelines for the definition of managed objects（GDMO）** 管理对象定义指南

**G2B（government to business）** 政府与企业

**G2C（government to citizen）** 政府与公众

**G2E（government to employee）** 政府与雇员

**G2G（government to government）** 政府与政府

# H

Hacker　黑客
Hacker attack　黑客攻击
Hacker ethic　黑客伦理观
Hacker jargon　黑客行话
hacker program　黑客程序
Hagelbarger code　黑格巴哥码
HAGO (have a good one)　就此搁笔
HAL (hardware abstraction layer)　硬件抽象层
half-bridge　半桥
half connection (HC)　半连接
half-duplex contention　半双工争用
half-duplex error protocol　半双工差错规程
half-gateway　半网关
half-session　半会话
Hamming distance　海明距离
Hamming net　海明网络
HAN (high-availability network)　高可用性网络
hand out red envelopes　发红包
hand shaking　握手
handshaking protocol　握手协议
hang a horse　挂马
hang horse program　挂马程序
hard error　硬差错
hardware abstraction layer (HAL)　硬件抽象层

hardware address　硬设备地址
hardware filtering　硬件过滤
hardware handshake　硬件握手信号
harmful out-of-band components　有害带外成分
harmonic response　谐波响应
hashnet interconnection　散列网络互连
haze computing　霾计算
HCDN (hybrid content delivery network)　混合内容发布网络
H-channel　H 信道
HCR (hybrid communication routing)　混合通信路由选择
HDB3 (high density bipolar of order 3 code)　三阶高密度双极性码
HDLC (high-level data link control)　高级数据链路控制
HDLC station　高级数据链路控制站
HDSL (high-bit-rate digital subscriber line)　高位速率数字用户线路
HDT (host digital terminal)　局用数字终端
HDX (half duplex)　半双工
headend unit　头端单元
head-on collision　冲突争用
hello packet　问候包
heterogeneous computer network　异构型计算机网络

heterogeneous convergence network 异构融合网络

heterogeneous multiplexing (HM) 异速复用

heterogeneous network 异构网络

heterogeneous network computing 异构网络计算

heterogeneous network convergence 异构网络融合

heterogeneous network environments 异构网络环境

heterogeneous network handover 异构网络切换

heterogeneous network interconnection 异构网络互连

heterogeneous network management 异构网络管理

heterogeneous network roaming 异网漫游

heterogeneous network system 异构网络系统

heterogeneous radio access technology 异构无线接入技术

heterogeneous wireless communication network 异构无线通信网络

heterogeneous wireless convergence network 异构无线融合网络

heterogeneous wireless network (HWN) 异构无线网络

heterogeneous wireless sensor network (HWSN) 异构无线传感器网络

heuristic routing 试探式路由选择

HF (high frequency) 高频

HFC (hybrid fiber coax) 混合光纤同轴

hidden node 隐藏节点

hidden node problem 隐藏节点问题

hidden terminal 隐藏终端

hidden terminal problem 隐藏终端问题

hierarchical computer network 层次计算机网络

hierarchical function network 层次泛函网络

hierarchical interconnection network 层次式互连网络

hierarchical mobile IP 分层移动 IP

hierarchical namespace 层次式名字空间

hierarchical network 分级网络

hierarchical network architecture (HNA) 分层网络体系结构

hierarchical network model 分层网络模型

hierarchical network model of semantic memory 分层语义网络

hierarchical network structure model 分层网络结构模式

hierarchical network node 分层网络节点

hierarchical networks routing 分层网络路由

hierarchical network structure 分层网络结构

hierarchical optical routing 分层光路由

hierarchical overlay network 分层网络

hierarchical multicast routing 分层组播路由

hierarchical role network model 分层

角色网络模型
hierarchical routing  分层路由选择
hierarchical routing tree  分层路由树
hierarchical routing network (HRN) 分级选路网
hierarchical satellite routing protocol 分级卫星路由协议
hierarchical signaling network  分层信令网
hierarchical state routing  分层状态路由
hierarchical synchronized network (HSN) 分级同步网络
hierarchical time stamp protocols (HTSP) 分层请求时间戳协议
high-availability network (HAN) 高可用性网络
high-bit-rate digital subscriber line (HDSL) 高位速率数字用户线路
high data rate (HDR)  高数据速率
high density bipolar of order 3 code (HDB3) 三阶高密度双极性码
high density bipolar (HDB) code 高密度双极性码
higher order path overhead monitor (HPOM) 高阶通道开销监视器
high layer communication protocol 高层通信协议
high layer compatibility (HLC) 高层兼容
high-level data link control (HDLC) 高级数据链路控制(协议)
high-level data link control adapter (HDLCA) 高级数据链路控制规程适配器
high-level data link control packet assembler/disassembler (HPAD) 高级数据链路控制分组装/拆设备
high-level data link control procedure 高级数据链路控制规程
high-level data link control station 高级数据链路控制站
high-level entity management standard (HEMS) 高级实体管理标准
high level protocol  高级协议
high performance parallel interface (HPPI) 高性能并行接口
high performance routing (HPR) 高性能路由选择
high rate packet data (HRPD) 高速分组数据
high speed burst communication 高速突发通信
high speed circuit service data (HSCSD) 高速电路数据业务
high-speed circuit-switched data (HSCSD) 高速电路交换数据
high-speed communication interface (HSCI) 高速通信接口
high speed downlink packet access (HSDPA) 高速下行链路分组接入
high-speed downlink shared channel (HS-DSCH) 高速下行链路共享信道
high-speed local network (HSLN) 高速局部网
high speed OFDM packet access (HSOPA) 高速 OFDM 分组接入
high speed packet access (HSPA) 高速分组接入
high speed packet switched data (HSPSD)

高速分组交换数据

high-speed serial interface（HSSI） 高速串行接口

high speed uplink packet access（HSUPA） 高速上行分组接入

hijacker 劫持者

HIN（hybrid integrated network） 混合综合网络

HIPS（host intrusion prevention system） 主机入侵预防系统

HLS（HTTP live streaming） HTTP直播流媒体

HMMA（hypermedia management architecture） 超媒体管理框架

HMMP（hypermedia management protocol） 超媒体管理协议

HMMS（hypermedia management schema） 超媒体管理机制

HMOM（hypermedia object manager） 超媒体对象管理者

HMSC（home mobile switching center） 归属移动交换中心

HNA（hierarchical network architecture） 分层网络体系结构

holddown timer 抑制计时器

holy war 圣战，网上争论

home location register（HLR） 归属位置寄存器

home mobile switching center（HMSC） 归属移动交换中心

home network 家庭网络

home network prefix 本地网络前缀

home office 家庭办公

home page 主页

HomePlug Powerline Alliance（HPA） 家庭插电联盟

home public land mobile network 本地公共陆地移动网

Home Radio-frequency Working Group（HRWG） 家用射频工作组

HomeRF 家庭射频，家用无线连接标准

homogeneous computer network 同构型计算机网络

homogeneous network 同构网络

homogeneous network environments 同构网络环境

honeynet skill 蜜网技术

honeypot skill 蜜罐技术

hop-by-hop route 逐跳路由

hop count 节点数，跳数

hop count limit 跨跃数限制

hop limit 跳数限制

HOPS（host proximity service） 邻近主机服务

horizontal link 水平链路

host digital terminal（HDT） 局用数字终端

hosted software 托管软件

host file 主机文件

host gateway 主机网关

host ID 宿主标识符

host interface 主机接口

host intrusion detection system（HIDS） 主机入侵检测系统

host intrusion prevention system（HIPS） 主机入侵预防系统

host monitoring protocol（HMP） 主机监督协议

host name 主机名

host node 主机节点

host number 主机号
host proximity service (HOPS) 邻近主机服务
host rental 主机租用
host route 主机路由
host system 宿主系统
host table 宿主表
host timed out 主机超时
host-to-host protocol 主机到主机协议
host unreachable 主机不可访
hot fix area 热修补区
hot image 热图像
hot list 热表,热门列表
hot potato routing 热土豆式路由选择
hot spot 热区,热点
hot standby router protocol (HSRP) 热备份路由器协议
hot text 热文本
hot zone 可用区,热区
HPA (HomePlug Powerline Alliance) 家庭插电联盟
HPAD (high-level data link control packet assembler/disassembler) 高级数据链路控制分组装/拆设备
HPPI (high performance parallel interface) 高性能并行接口
HPR (high performance routing) 高性能路由选择
HRC (hybrid ring control) 混合环控制
HRC cycle HRC 轮转环
HRWG (home radio-frequency working group) 家用射频工作组
HRN (hierarchical routing network) 分级选路网
HRPD (high rate packet data) 高速分组数据

HSCI (high-speed communication interface) 高速通信接口
HSCSD (high speed circuit service data) 高速电路数据业务
HSCSD (high-speed circuit-switched data) 高速电路交换数据
HSDPA (high speed downlink packet access) 高速下行链路分组接入
HS-DSCH (high-speed downlink shared channel) 高速下行链路共享信道
HSLN (high-speed local network) 高速局部网
HSN (hierarchical synchronized network) 分级同步网络
HSOPA (high speed OFDM packet access) 高速 OFDM 分组接入,高速正交频分复用分组接入
HSPA (high speed packet access) 高速分组接入
HS-PDSCH (high-speed physical downlink shared channel) 高速物理下行共享信道
HSPSD (high speed packet switched data) 高速分组交换数据
HSRP (hot standby router protocol) 热备份路由器协议
HSSI (high-speed serial interface) 高速串行接口
HSUPA (high speed uplink packet access) 高速上行分组接入
.htm HTML 文件名后缀
.html HTML 文件名后缀
HTML (hypertext mark-up language) 超文本标记语言

HTML table　超文本标记语言表
HTML validation service　HTML 验证服务
HTSP (hierarchical time stamp protocols)　分层请求时间戳协议
HTTP (hypertext transport protocol)　超文本传输协议
HTTP client　HTTP 客户端
HTTPD (hypertext transport protocol daemon)　超文本传输协议常驻程序
HTTP live streaming (HLS)　HTTP 直播流媒体
HTTP message　HTTP 报文
HTTP next generation (HTTP-NG)　下一代超文本传输协议
HTTP-NG (HTTP next generation)　下一代超文本传输协议
HTTP payload entity　HTTP 载荷实体
HTTP proxy　HTTP 代理
HTTP representation　HTTP 表示
HTTP request message　HTTP 请求报文
HTTP response message　HTTP 响应报文
HTTPS (hypertext transfer protocol over secure socket layer)　超文本传输安全协议
HTTP server　HTTP 服务程序
hub　集线器
hub board　集线器板
hub go-ahead　向内探询
hub server　集线器服务器
human flesh search　人肉搜索
human flesh search engine　人肉搜索引擎
HWN (heterogeneous wireless network)　异构无线网络
HWNN (hybrid wavelet neural network)　混合小波神经网络
hybrid access method　混合访问方法
hybrid active network　混合主动网
hybrid architecture for media streaming　混合流媒体结构
hybrid automatic repeat request (HARQ)　混合自动重发请求
hybrid Bayesian network　混合贝叶斯网络
hybrid cloud　混合云
hybrid communication　混合通信
hybrid communication model　混合通信模式
hybrid communication network　混合通信网
hybrid communication routine　混合通信路径
hybrid communication routing (HCR)　混合通信路由选择
hybrid content delivery network (HCDN)　混合内容发布网络
hybrid cube network　混合立方形网络
hybrid error control (HEC)　混合差错控制
hybrid fiber coax (HFC)　混合光纤同轴
hybrid fiber coaxial access network　混合光纤同轴接入网
hybrid fiber wireless system　光纤无线混合系统
hybrid integrated network (HIN)　混合综合网络
hybrid local network　混合局域网
hybrid division multiple access (xDMA)

混合多址
hybrid mobile network 混合移动网络
hybrid multiple access protocol 混合多址协议
hybrid network 混合网络
hybrid neural network 混合神经网络
hybrid neural network model 混合神经网络模型
hybrid optical switching 混合光交换
hybrid P2P on-demand streaming media system 混合式 P2P 点播流媒体系统
hybrid radio network 混合无线网
hybrid ring control (HRC) 混合环控制
hybrid routing 混合式路由选择
hybrid routing algorithm 混合路由算法
hybrid routing model 混合路由模型
hybrid routing protocols 混合式路由选择协议
hybrid spread spectrum communication 混合扩频通信
hybrid spread spectrum modulation 混合扩谱调制
hybrid star topology 混合星形拓扑
hybrid stream media distribution model 混合流媒体分发模型
hybrid switching 混合交换
hybrid switching optical network 混合交换光网络
hybrid telecommunication 混合远程通信
hybrid topology 混合拓扑
hybrid wavelet neural network (HWNN) 混合小波神经网络
hybrid wireless mesh network 混合无线网状网
hybrid wireless mesh protocol 混合无线

网状网协议
hybrid wireless network 混合无线网络
hyperaccess 超级访问
hyperlink 超链接
hypermedia 超媒体
hypermedia management architecture (HMMA) 超媒体管理框架
hypermedia management protocol (HMMP) 超媒体管理协议
hypermedia management schema (HMMS) 超媒体管理机制
hypermedia object manager (HMOM) 超媒体对象管理者
hypermedia time-based structuring language (HyTime) 时基超媒体结构化语言
hyperrectangular netware 超矩形网络
hyperspace 超空间
hypertext 超文本
hypertext link 超文本链接
hypertext mark-up language (HTML) 超文本标记语言
hypertext mark-up language 5 (HTML 5) 超文本标记语言第 5 版
hypertext preprocessor (PHP) 超文本预处理器
hypertext technology 超文本技术
hypertext transport protocol (HTTP) 超文本传输协议
hypertext transfer protocol over secure socket layer (HTTPS) 超文本传输安全协议
hypertext transport protocol daemon (HTTPD) 超文本传输协议常驻程序
hypothetical reference connection (HRX) 假设参考连接

**hypothetical reference digital path at 64 kbps**　64kbps 的假设参考数字通道

**hypothetical reference digital section**（HRDS）　假设参考数字段

**HyTime (hypermedia time-based structuring language)**　时基超媒体结构化语言

# I

IaaS (infrastructure as a service) 基础设施即服务
IAB (Internet Architecture Board) 因特网结构委员会
IAM (inverse multiplexing over ATM) ATM 上的反向复用
IANA (Internet Assigned Numbers Authority) 因特网分配号码管理局
IANAL (I am not a lawyer) 我不是律师
IAP (Internet Access Provider) 因特网接入提供商
IAP (intrusion alert protocol) 入侵警报协议
IAS (Internet access service) 因特网接入服务
IAS (intruder alarm system) 入侵报警系统
IBANT (International Borker Antivirus Network Technology Alliance) 国际白客网络安全技术联盟
IBGP (internal border gateway protocol) 内部边界网关协议
IBM token ring　IBM 令牌环
IBM token ring network　IBM 令牌环网
ICA (International Communication Association) 国际通信协会
ICANN (Internet Corporation for Assigned Names and Numbers) 因特网名称与数字地址分配机构
ICCB (Internet Configuration Control Board) 因特网结构控制研究会
ICE (information and content exchange) 信息与内容交换(协议)
ICI (interchannel interference) 信道间干扰
ICI (interface control information) 接口控制信息
ICI (inter carrier interface) 载波间接口
ICMRA (inner cluster multiple-hop routing algorithm) 基于转发意愿的簇内多跳路由算法
ICMP (Internet control message protocol) 因特网控制信息协议
ICMP router discovery protocol (IRDP) ICMP 路由器发现协议
ICP (Internet Content Provider) 因特网内容提供商
ICP (internet cache protocol) 互联网高速缓存协议
ICQ (I seek you) 网络寻呼机
ICS (nternet connection sharing) 因特网连接共享服务
ID (identifier) 标识符

I-D (Internet-draft) 因特网草案

IDC (internet data central) 互联网数据中心

IDC (internet database connector) 互联网数据库连接器

ideal routing algorithm 理想路由算法

identity association (IA) 身份关联

identity link 一致性连接

IDF (International DOI Foundation) 国际数字对象识别号基金会

IDLC (integrated digital loop carrier) 综合数字环路载波

idle communication mode 空闲通信方式

idle RQ 空闲RQ(协议)

IDN (integrated digital network) 综合数字网

IDN (isolated data network) 独立数据网络

IDNs (international domain names) 国际域名

IDRP (inter-domain routing protocol) 域间路由选择协议

IDS (intrusion detection system) 入侵检测系统

IDXP (intrusion detection exchange protocol) 入侵检测交换协议

IE (information extraction) 信息抽取

IE (information element) 信元

IEC (International Electrotechnical Commission) 国际电工技术委员会

IEEE (Institute of Electrical and Electronic Engineers) 电气与电子工程师学会

IEEE 802 committee IEEE802 委员会

IEEE 802 standards IEEE802 标准

IEEE 802.1 standard IEEE802.1 标准

IEEE 802.1x standard IEEE802.1x 标准,定义了基于端口的网络访问控制协议

IEEE 802.11 standard IEEE802.11 标准,定义了无线局域网系列标准

IEEE 802.14 standard IEEE802.14 标准,用于协调混合光纤同轴(HFC)网络的前端和用户站点间数据通信的协议

IEEE 802.15 standard IEEE802.15 标准,无线个人区域网络(WPAN)标准

IEEE 802.16 standard IEEE802.16 标准,宽带无线城域网标准(WiMAX)

IEEE 802.2 standard IEEE802.2 标准,局域网逻辑链路控制(LLC)子层的标准

IEEE 802.20 standard IEEE802.20 标准,移动宽带接入技术(MBWR)标准

IEEE 802.3 standard IEEE802.3 标准,局域网物理接线标准

IEEE 802.4 standard IEEE802.4 标准,令牌传送总线式局域网的物理层和介质访问控制(MAC)子层标准

IEEE 802.5 standard IEEE802.5 标准,令牌环式局域网的物理层和介质存取控制(MAC)子层的标准

IEEE 802.6 standard IEEE802.6 标准,城域网访问法和物理层技术规范的标准

IEEE 802.9 standard IEEE802.9 标准,IEEE 综合数据语音网络小组(IDVNG)关于将语音、数据和视频图像综合入局域网和 ISDN(综合业务

数字网)的标准

IEN (Internet engineering note) 因特网工程备忘录

IEPG (Internet engineering and planning group) 因特网工程和计划组

IESG (Internet Engineering Steering Group) 因特网工程管理组

IETF (Internet Engineering Fask Force) 因特网工程任务组

IFHO (inter-frequency handoff) 频间切换

IFMP (Ipsilon flow management protocol) Ipsilon 专用流量管理协议

IFRB (International Frequency Registration Board) 国际频率登记局

IGA (in-game advertising) 网游植入广告

IGMP (internet group management protocol) 互联网组管理协议

IGMP Proxy (internet group management protocol proxy) 互联网组管理协议代理

IGMP Snooping (internet group management protocol snooping) 互联网组管理协议监听

IGP (interior gateway protocol) 内部网关协议

IGRP (Internet gateway routing protocol) 互联网网关路由协议

IGRP (interior gateway routing protocol) 内部网关路由协议

IIN (integrated intelligent network) 综合智能网

IIOP (Internet inter-ORB protocol) 因特网内部对象请求代理协议

IIS (Internet information server) 因特网信息服务器

IISP (interim inter switch protocol) 临时交换机间信令协议

IKE (Internet key exchange) 因特网密钥交换

i-key protocol (i-KP) 面向电子商务的支付协议

i-KP (i-KP protocol) 面向电子商务的支付协议

ILEC (incumbent local exchange carrier) 本地交换运营商

ILMI (integrated link management interface) 集成链路管理接口

ILMI (interim local management interface) 临时本地管理接口

IMAP (Internet message access protocol) 因特网信息访问协议

IMAP (interactive mail access protocol) 交互式邮件访问协议

IMC (Internet Mail Consortium) 因特网邮件联盟

IMHO (in my humble opinion) 以鄙人之见,恕我直言

IMO (in my opinion) 依我之见

IMP-IMP protocol 接口机-接口机协议

implicit congestion notification 隐式拥塞挤通知

IMP-to-host interface 接口信息处理机到主机的接口

IMP-to-host protocol 接口信息处理机到主机的协议

improved artificial neural network 改

进人工神经网络
IMSI (International Mobile Sbscriber Identity) 国际移动用户标志
IN (intelligent network) 智能网
inactive link 非活动链
INAP (intelligent network application protocol) 智能网络应用协议
INAS (integrated network active system) 综合网络激活系统
in-band signaling 带内信令
inbound link 导入链接
inbound path 入站路径
incumbent local exchange carrier (ILEC) 本地交换运营商
independent WLAN 独立式无线局域网
in-dialing network 互联拨号网
indirectly-coupled system 间接耦合系统
indirect network 间接网络
indoor broadband wireless network (IBWN) 室内宽带无线网络
indoor covering system 室内覆盖系统
industrial heterogeneous network 工业异构网络
industrial, scientiflc and medical (ISM) 工业、科学和医用
industry cloud 行业云
Industry Standard Architecture (ISA) 工业标准体系结构
industry ubiquitous wireless sensor network (IUWSN) 工业泛在传感网络
INE (intermediate network element) 中间网络元素

INFM (intelligent network functional model) 智能网功能模型
Infobahn (information Autobahn) 信息巴恩,信息高速公路
information and content exchange (ICE) 信息与内容交换(协议)
information appliance (IA) 信息电器
information architecture (IA) 信息架构
information bearer channel 带控制信息的信道
information behavior 信息行为
information bit 信息位
information element (IE) 信元
information filtering system 信息过滤系统
information infrastructures 信息基础设施
information island 信息孤岛
information island integration 信息孤岛集成
information island phenomenon 信息孤岛现象
information network 信息网
information network dissemination power protection rule 信息网络传播权保护条例
information operation 信息作战
information outlet 信息输出口
information packet 信息包
information path 信息通路
information payload 信息净负荷
information pull 信息拉取
information push 信息推送
information push-pull 信息推拉

information resource sharing network 信息资源共享网络
information retrieval system 信息检索系统
information sharing network 信息共享网络
information superhighway 信息高速公路
information system outsourcing 信息系统外包
information technology outsourcing 信息技术外包
Infrared Data Association (IrDA) 红外线数字协会
infrared link access protocol (IrLAP) 红外连接访问协议
infrared link management protocol (IrLMP) 红外连接管理协议
infrared physical layer link specification (IrPHY) 红外物理层连接规范
infrastructure as a service (IaaS) 基础设施即服务
in-game advertising (IGA) 网游植入广告
initial cell rate (ICR) 初始信元速率
initial chaining value (ICV) 初始链接值
initial connection protocol 初始连接协议
initiating LU (ILU) 初始逻辑单元
injection attack 注入式攻击
inker 印客
inner cluster multiple-hop routing algorithm (ICMRA) 基于转发意愿的簇内多跳路由算法

INM (integrated network management) 综合网络管理
INMS (integrated network management system) 综合网络管理系统
in my humble opinion (IMHO) 以鄙人之见，恕我直言
in my opinion (IMO) 依我之见
INN (intermediate network node) 中间网络节点
I-NNI (interior-network to network interface) 内部网络-网络接口
INON (intelligent node overlay network) 智能节点重叠网
in-plant system 在场系统
inquiry and communication system 查询与通信系统
inquiry and response system 询问与应答系统
inquiry reply 询问应答
inquiry/response 查询和响应
inquiry/response communication 查询和响应通信
inquiry/response operation 查询和响应操作
INS (intelligent network service) 智能网络服务
in service monitoring (ISM) 带业务监测
inside global address 内部全局地址
inside link 内部链接，内链
inside local address 内部本地地址
instant message (IM) 即时消息
Institute of Electrical and Electronic Engineers (IEEE) 电气与电子工程师学会

integrated access media gateway 综合接入媒体网关
integrated broadband access system 综合宽带接入系统
integrated broadband communication 综合宽带通信
integrated broadband communication network 综合宽带通信网
integrated broadband network 综合宽带网
integrated broadband system 综合宽带系统
integrated digital communication network 综合数字通信网
integrated digital enhanced network 综合数字增强型网络
integrated digital loop carrier (IDLC) 综合数字环路载波
integrated digital network (IDN) 综合数字网
integrated intelligent network (IIN) 综合智能网
integrated Internet service 集成化因特网服务
integrated IS-IS 综合IS-IS，综合中间系统到中间系统协议
integrated link management interface (ILMI) 集成链路管理接口
integrated marketing 整合营销
integrated network active system (INAS) 综合网络激活系统
integrated network management (INM) 综合网络管理
integrated network management agent 综合网络管理代理
integrated network management system (INMS) 综合网络管理系统
integrated network processor 集中网络处理机
integrated network security protection 一体化网络安全防护
integrated open hypermedia (IOH) 集成开放超媒体
integrated service (IntServ) 集成服务
integrated service access gateway 综合业务接入网关
integrated service access network 综合业务接入网
integrated service access point 综合业务接入访问点
integrated service control point (ISCP) 综合业务控制点
integrated service data point (ISDP) 综合业务数据点
integrated service digital network (ISDN) 综合业务数字网
integrated service generation environment point (ISGEP) 综合业务生成环境点
integrated service line unit (ISLU) 综合业务线路单元
integrated service local network (ISLN) 综合业务局部网
integrated service management point (ISMP) 综合业务管理点
integrated service management access point (ISMAP) 综合业务管理接入点
integrated service switching point (ISSP) 综合业务交换点

integrated service unit (ISU) 综合业务单元
integrated services wireless access network 综合业务无线接入网络
integrated setuives network (ISN) 综合业务网
integrated switch/router (ISR) 集成交换/路由器,交换路由一体机
integrated telecommunication services 综合电信业务
integrated virtual private network 综合虚拟专用网
integrated voice data local area network (IVD-LAN) 综合声音数据局域网
intelligent automatically switched optical network 智能化自动交换光网络
intelligent community information system 智能社区信息系统
intelligent domain name resolution 智能域名解析
intelligent flexible spatial system 智慧弹性空间系统
intelligent hub 智能集线器
intelligent information push-pull 智能信息推拉
intelligent integrated information network 智能综合信息网络
intelligent interfacing 智能接口
intelligent multimedia 智能多媒体
intelligent network (IN) 智能网
intelligent network agent 智能网络代理
intelligent network application protocol (INAP) 智能网络应用协议
intelligent network architecture 智能网体系结构
intelligent network functional model (INFM) 智能网功能模型
intelligent network service (INS) 智能网络服务
intelligent node overlay network (INON) 智能节点重叠网
intelligent peripheral (IP) 智能外设
intelligent residential district 智能小区
intelligent spatial decision 智能空间决策
Intelligent spatial decision support system 智能空间决策支持系统
intelligent spatial information service 智能空间信息服务
interaction network 互动网络
interactive advertising 交互式广告
interactive magazine 互动杂志
interactive mail access protocol (IMAP) 交互式邮件访问协议
interactive network 交互网络
interactive network adapter 交互网络适配器
interactive personality television (IPTV) 交互式网络电视
interactive smart grid 互动(智能)电网
Inter-area path 区域间路径
inter-area route 区域间路由
interband transmission 带间传输
inter carrier interface (ICI) 载波间接口
interchange node 交换节点
interchange transmission group (TG)

交换传输组

interchannel interference 信道间干扰
intercommunicating system 内部通信系统
inter-dimensional routing 维间路由选择
inter-domain authentication 域间认证
inter-domain multicast 域间组播
inter-domain multicast service 域间组播服务
Inter-domain negotiation 域间协商
inter-domain QoS 域间服务质量
inter-domain route 域间路由
Inter-domain route optimization 域间路由优化
inter-domain route system 域间路由系统
inter-domain routing protocol (IDRP) 域间路由选择协议
inter-domain signaling 域间信令
inter-domain trust 域间信任
inter-domain direct trust 域间直接信任
inter-domain reputation trust 域间声望信任
inter exchange link (IXL) 局间链路
interface control information (ICI) 接口控制信息
interface data unit (IDU) 接口数据单元
interface message processor (IMP) 接口信息处理机,接口机
interface node 接口节点
interface processor 接口处理机
interface rate 接口速率

interface service node 接口服务节点
interframe coding 帧间编码
interframe comparison 帧间比较
interframe compression 帧间压缩
interframe continuity 帧间连续性
Interframe forecast 帧间预测
Interframe gap 帧间距
interframe spacing 帧间间隔
interframe stabilization 帧间稳定
inter-frequency handoff (IFHO) 频间切换
interim inter switch protocol (IISP) 临时交换机间信令协议
interim local management interface (ILMI) 临时本地管理接口
interior gateway 内部网关
interior gateway protocol (IGP) 内部网关协议
interior gateway routing protocol (IGRP) 内部网关路由协议
interior-network to network interface (I-NNI) 内部网络-网络接口
interior switch 内部交换器
interior switching link (ISL) 交换链路内协议
intermediate host node 中间主机节点
intermediate network node 中间机网络节点
intermediate node 中间节点
intermediate protocol driver (IPD) 中间驱动程序
intermediate router 中间路由器
intermediate routing function 中间路由选择功能
intermediate routing network 中间路

由网络

intermediate routing node (IRN) 中间路由选择节点

intermediate network node (INN) 中间网络节点

intermediate network element (INE) 中间网络元素

intermediate routing network 中间路由网络

interim network organization 中间网络组织

intermediate signaling network identification (ISNI) 中间信令网络识别

intermediate session routing (ISR) 交互会话路由选择

intermediate system (IS) 中间[中介]系统

intermediate system-intermediate system (IS-IS) 中间系统到中间系统

Intermediate system routing exchange protocol 中间系统路由交换协议

intermediate TCAM node 中间TCAM节点

internal border gateway protocol (IBGP) 内部边界网关协议

internal gateway protocol 内部网关协议

internal gateway routing protocol 内部网关路由协议

internal cloud 内云

internal home network 室内网络

internal network detection 内部网络检测

internal reachable address 内部可达地址

internal router 区域内路由器

International access code 国际接入码

International Borker Antivirus Network Technology Alliance (IBANT) 国际白客网络安全技术联盟

International code designator (ICD) 国际代码指派者

International DOI Foundation (IDF) 国际数字对象识别号基金会

International domain names (IDNs) 国际域名

International domain name dispute resolution mechanism 国际统一域名争议解决机制

International Frequency Registration Board (IFRB) 国际频率登记局

International Mobile Subscriber Identity (IMSI) 国际移动用户标志

International mobile telecommunication 2000 (IMT－2000) 国际移动通信2000

International Multimedia Telecommunication Consortium (IMTC) 国际多媒体通信协会

International Network Working Group (INWG) 国际网络工作组

International Organization for Standardization Open Systems Interconnection model (ISO/OSI model) 国际标准化组织/开放式系统互联模型

International prefix 国际前缀

International Science & Education Researcher Association (ISER) 国际科教研究者协会

International Softswitch Consortium (ISC) 国际软交换协会
International Standardization Organization (ISO) 国际标准化组织
International Telecommunication Union (ITU) 国际电信联盟
International Telecommunication Union-Telecommunication Standardization Section (ITU-TSS) 国际电信联盟-电信标准化部门
International top level domain names (iTLD, iTDs) 国际顶级域名
International virtual private network 国际虚拟专用网
internet 互联网
Internet 因特网
Internet Access Provider (IAP) 因特网接入提供商
Internet access service (IAS) 因特网接入服务
Internet account 因特网账号
internet adaptive connection 互联网自适应互联
Internet address 因特网地址
internet advertisement media 网络广告媒体
Internet application middleware 因特网应用中间件
Internet Architecture Board (IAB) 因特网结构委员会
Internet Assigned Numbers Authority (IANA) 因特网分配号码管理局
internet bank 网络银行
internet broadcasting 互联网广播
internet cache 互联网高速缓存

internet cache protocol (ICP) 互联网高速缓存协议
Internet Configuration Control Board (ICCB) 因特网结构控制研究会
internet congestion control 互联网拥塞控制
internet connection firewall 互联网连接防火墙
Internet connection sharing (ICS) 因特网连接共享服务
Internet content filtering 因特网内容过滤
Internet Content Provider (ICP) 因特网内容提供商
internet control gateway 互联网控制网关
Internet control message protocol (ICMP) 因特网控制信息协议
Internet Corporation for Assigned Names and Numbers (ICANN) 因特网名称与数字地址分配机构
internet crisis 互联网危机
internet crisis management 互联网危机管理
internet database 互联网数据库
internet database connector (IDC) 互联网数据库连接器
internet data central (IDC) 互联网数据中心
internet digital subscriber line (IDSL) 互联网数字用户线路
Internet-draft (I-D) 因特网草案
internet eco crisis 网络生态危机
internet economy 网络经济
internet encyclopedia (Interpedia) 网

络百科全书

Internet Engineering and Planning Group (IEPG)　因特网工程和计划组

Internet engineering note (IEN)　因特网工程备忘录

Internet Engineering Steering Group (IESG)　因特网工程管理组

Internet Engineering Task Force (IETF)　因特网工程任务组

internet exchange architecture (IXA)　互联网交换架构

Internet Exchange Centre (IXC)　因特网交换中心

internet finance　互联网金融

internet firewall　互联网防火墙

internet gambling　互联网赌博

internet gateway routing protocol (IGRP)　互联网网关路由协议

internet group management protocol (IGMP)　互联网组管理协议

internet group management protocol proxy (IGMP Proxy)　互联网组管理协议代理

internet group management protocol snooping (IGMP Snooping)　互联网组管理协议监听

internet header length (IHL)　互联网标题长度

Internet information server (IIS)　因特网信息服务器

Internet information service　因特网信息服务

Internet inter-ORB protocol (IIOP)　因特网内部对象请求代理协议

Internet key exchange (IKE)　因特网密钥交换

Internet Mail Consortium (IMC)　因特网邮件联盟

internet marketing　互联网营销

Internet message access protocol (IMAP)　因特网信息访问协议

internet MLM　网络传销

internet of things (IOT)　物联网

internet of things foundation　物联网基础

internet of things engineering　物联网工程

Internet PCA Registration Authority (IPRA)　因特网公共密码学注册机构

Internet phone　因特网电话

Internet phone gateway　因特网电话网关

Internet Policy Registration Authority (IPRA)　因特网政策注册管理机构

Internet Presence Provider (IPP)　因特网平台提供商

internet protocol (IP)　网际协议

internet protocol multicasting (IPM)　IP多(路广)播

internet protocol (IP) network　IP网

internet protocol next generation (IPng)　下一代IP协议

Internet Protocol Security (IPSec)　因特网协议安全性

internet public opinion　网络舆论,网络舆情

internet public opinion supervision　网络舆情监督

internet reference model　互联网参考

模型

Internet registry (IR) 因特网登记处

Internet relay chat (IRC) 因特网中继对话

Internet remote control network 因特网远程控制网络

Internet Research Steering Group (IRSG) 因特网研究管理组

Internet Research Task Force (IRTF) 因特网研究任务组

internet new media 网络新媒体

internet rich media 网络富媒体

internet security 互联网安全性

internet security threat report 互联网安全威胁报告

Internet security association and key management protocol (ISAKMP) 因特网安全连接和密钥管理协议

internet server application program interface (ISAPI) 互联网服务器应用程序接口

Internet service manager (ISM) 因特网服务管理器

Internet Service Provider (ISP) 因特网服务提供商

Internet shopping network (ISN) 因特网购物网

internet short message gateway (ISMG) 短信网关

internet small computer system interface (iSCSI) 互联网小型计算机系统接口

internet social capital 网络社会资本

internet social contact 网络社交

Internet social engineering attack 因特网社会工程攻击

Internet Society (ISOC) 因特网协会

Internet Society of China (ISC) 中国互联网协会

Internet Software Consortium (ISC) 因特网软件联盟

Internet Streaming Media Alliance (ISMA) 网络流媒体联盟

internet streaming media broadcast 网络流媒体直播

internet systems engineering 互联网系统工程

internet sovereignty 网络主权

internet storage name service (iSNS) 互联网存储命名服务

Internet talk radio (ITR) 因特网无线电对话

Internet telephone 因特网电话

Internet telephone gateway 因特网电话网关

internet traffic engineering 互联网流量工程

Internet transport service 因特网传输服务

internet trust crisis 网络信任危机

internetworking 网际互联

internetworking bridge 网际桥接器

internetwork data separator 网际数据隔离器

internetworking protocol 网际互连协议

internetworking technologies 网络互连技术

internetwork packet 网际分组,互联网数据包

internetwork packet exchange(IPX) 互联网数据包交换

internetwork packet exchange protocol 互联网数据包交换协议

internetwork packet exchange/sequenced packet exchange(IPX/SPX) 互联网数据包交换/顺序包交换

internet worm 互联网蠕虫

Internet 2 第二代因特网

internet+ 互联网+

InterNIC(Internet network information center) 因特网网络信息中心

internodal destination queue 节点间目的地队列

internodal sequence number synchronization 节点间顺序号同步

internode routing 节点间路由选择

Interpedia(internet encyclopedia) 网络百科全书

interpersonal messaging service 人际消息传递业务

interprocess communication(IPC) 进程间通信

inter-route switching 路由间交换

inter switching system interface(ISSI) 交换系统间接口

intersymbol dependence 码间相关性

intersymbol interference(ISI) 符号间干扰

interworking function(IWF) 互通功能

interworking network 互通网络

interworking unit 互通设备

intrabuilding network cable 楼内网络电缆

intradimensional routing 维内路由选择

intra frame coding 帧内编码

intraframe compression 帧内压缩

Intranet 内联网

Intranet security 内联网安全

intranode routing 节点内路由选择

intra-route switching 路由内交换

intra-site automatic tunnel addressing protocol(ISATAP) 内部站点自动隧道寻址协议

intruder alarm system(IAS) 入侵报警系统

intrusion alert protocol(IAP) 入侵警报协议

intrusion detection exchange format(IDEF) 入侵检测交换格式

intrusion detection exchange protocol(IDXP) 入侵检测交换协议

intrusion detection message exchange format(IDMEF) 入侵检测消息交换格式

intrusion detection system(IDS) 入侵检测系统

intrusion detection working group(IDWG) 入侵检测工作组

intrusion prevention system(IDS) 入侵预防系统

intrusion response system(IRS) 入侵响应系统

IntServ(integrated service) 集成服务

invalid time stamp 无效时间戳

inverse multiplexing over ATM(IAM) ATM上的反向复用

INWG(International Network Working

Group) 国际网络工作组
IOT (internet of things) 物联网
IP (Internet protocol) 网际协议
IP access  IP 接入
iPad  苹果平板计算机
IP address  IP 地址
IP address mask  IP 地址屏蔽
IP address prefix  IP 地址前缀
IP address spoofing  IP 地址欺骗
IP address spoofing attack  IP 地址欺骗攻击
IPAH (IP authentication header)  IP 验证报头
IP authentication header (IPAH)  IP 验证报头
IP-based SAN  基于 IP 存储区域网
IP broadcast address  IP 广播地址
IP broadcast network  IP 广播网络
IPC (interprocess communication)  进程间通信
IP-CAN (IP connectivity access network)  IP 连接访问网络
IP connectivity access network (IP-CAN)  IP 连接访问网络
IP core interface layer  IP 核心接口层
IP data packet  IP 数据包
IP datagram  IP 数据报
IP encapsulating security payload (IPESP)  IP 封装安全载荷
IPESP (IP encapsulating security payload)  IP 封装安全载荷
IP Fax  IP 传真
IP floating address  IP 地址漂移
IP gateway  IP 网关
IP group multicast  IP 组播

iPhone  苹果手机
IP layer multicast  IP 层组播
IP layer security  IP 层安全
IPLS (IP-only LAN-like service)  只支持 IP 的局域网业务
IPM (internet protocol multicasting)  IP 多(路广)播
IP mobile phone  IP 移动电话
IP multicast  IP 组播
IP multicast member access control  IP 组播组成员访问控制
IP multicast service  IP 组播业务
IP multicast technology  IP 组播技术
IP multimedia public identity  IP 多媒体公用识别码
IP multimedia services  IP 多媒体业务
IP multimedia streaming services  IP 多媒体流业务
IP multimedia subsystem (IMS)  IP 多媒体子系统
IP multimedia network subsystem  IP 多媒体网络子系统
IP network  IP 网
IP network layer security  IP 网络层安全
IP network management layer  IP 网络管理层
IP network phone  IP 网络电话
IPng (internet protocol next generation)  下一代 IP 协议
IPOA (IP over ATM)  ATM 上的 IP 传输
IP-only LAN-like service (IPLS)  只支持 IP 的局域网业务
IPOS (IP over SDH)  SDH 上的 IP 传

输,同步数字系列上的 IP 传输

IP over ATM (IPOA) ATM 上的 IP 传输

IP over SDH (IPOS) SDH 上的 IP 传输,同步数字系列上的 IP 传输

IP over WDM (IPOW) WDM 上的 IP 传输,波分复用上的 IP 传输

IPOW (IP over WDM) WDM 上的 IP 传输,波分复用上的 IP 传输

IPP (Internet Presence Provider) 因特网平台提供商

IP packet buffer memory IP 包缓存

IP packet delay variation IP 包时延变化

IP packet filter firewall IP 包过滤防火墙

IP packet loss ratio IP 包丢失率

IP packet multicast IP 包组播

IP packet over SDH 基于 SDH 的 IP 分组,基于同步数字系列上的 IP 分组

IP-PBX (IP private branch exchange) IP 专用交换分机

IP phone IP 电话

IP private branch exchange (IP-PBX) IP 专用交换分机

IP public network IP 公用网

IPR (isolated pacing response) 隔离整步响应

IPRA (Internet Policy Registration Authority) 因特网政策注册管理机构

IPRA (Internet PCA Registration Authority) 因特网公共密码学注册机构

IP registered address IP 注册地址

IPS (intrusion prevention system) 入侵预防系统

IPSec (Internet Protocol Security) 因特网协议安全性

Ipsilon flow management protocol (IFMP) Ipsilon 专用流量管理协议

IPSS (International Packet Switching Service) 国际数据包交换服务机构

IP storage network IP 存储网络

IP switch IP 交换

IP telephony network IP 电话网

IPTV (interactive personality television) 交互式网络电视

IPTV middleware IPTV 中间件

IPTV service IPTV 业务

IP-UMTS (all IP universal mobile telecommunication system) 全 IP 通用移动通信系统

IP unnumbered address IP 无编号地址

IP video phone IP 可视电话

IPv4 address IPv4 地址

IPv4 multicast address IPv4 组播地址

IPv6 address IPv6 地址

IPv6 multicast address IPv6 组播地址

IPX (internetwork packet exchange) 互联网数据包交换

IPX external network number IPX 外部网络数

IPX internal network number IPX 内部网络数

IPX/SPX (internetwork packet exchange/sequenced packet exchange) 互联网数据包交换/顺序包交换

IR (Internet registry) 因特网登记处

**IRC**（Internet relay chat） 因特网中继对话

**IrDA**（Infrared Data Association） 红外线数字协会

**IRDP**（ICMP router discovery protocol） ICMP 路由器发现协议

**IrLAP**（infrared link access protocol） 红外连接访问协议

**IrLMP**（infrared link management protocol） 红外连接管理协议

**IRN**（intermediate routing node） 中间路由选择节点

**IrPHY**（infrared physical layer link specification） 红外物理层连接规范

**IRS**（intrusion response system） 入侵响应系统

**IRSG**（Internet Research Steering Group） 因特网研究管理组

**IRTF**（Internet Research Task Force） 因特网研究任务组

**ISA**（Industry Standard Architecture） 工业标准体系结构

**IS**（intermediate system） 中间系统，中介系统

**ISAKMP**（Internet security association and key management protocol） 因特网安全连接和密钥管理协议

**ISAPI**（internet server application program interface） 互联网服务器应用程序接口

**ISATAP**（intra-site automatic tunnel addressing protocol） 内部站点自动隧道寻址协议

**ISATAP address** ISATAP 地址

**ISATAP tunneling** ISATAP 隧道

**ISC**（Internet Society of China） 中国互联网协会

**ISC**（Internet Software Consortium） 因特网软件联盟

**ISC**（International Softswitch Consortium） 国际软交换协会

**ISCCC**（China Information Security Certification Center） 中国信息安全认证中心

**ISCP**（integrated service control point） 综合业务控制点

**iSCSI**（internet small computer system interface） 互联网小型计算机系统接口

**ISDN**（integrated service digital network） 综合业务数字网

**ISDN terminal** 综合业务数字网终端

**ISDN user part**（ISUP） 综合业务数字网用户部分

**ISDP**（integrated service data point） 综合业务数据点

**I seek you**（ICQ） 网络寻呼机

**ISGEP**（integrated service generation environment point） 综合业务生成环境点

**ISER**（International Science & Education Researcher Association） 国际科教研究者协会

**IS-IS**（intermediate system-intermediate system） 中间系统到中间系统

**ISL**（interior switching link） 交换链路内协议

**ISLU**（integrated service line unit） 综合业务线路单元

**ISM**（in service monitoring） 带业务

监测

ISM（Internet service manager） 因特网服务管理器

ISMA（Internet Streaming Media Alliance） 网络流媒体联盟

ISMAP（integrated service management access point） 综合业务管理接入点

ISMG（internet short message gateway） 短信网关

ISMP（integrated service management point） 综合业务管理点

ISN（Internet shopping network） 因特网购物网

ISN（integrated setuives network） 综合业务网

ISNI（intermediate signaling network identification） 中间信令网络识别

ISO（International Standardization Organization） 国际标准化组织

ISOC（Internet Society） 因特网协会

isochronous network 等时网络，同步网络

isochronous service 等时服务，等时服务

isochronous transmission 等时传输，同步传输

IsoEthernet 等时以太网

isolated adaptive routing 孤立自适应路由选择

isolated bonding network 隔离连接网络

isolated data network（IDN） 独立数据网络

Isolated network 独立网络

isolated pacing response（IPR） 隔离步响应

isolated power network 独立电力网络

ISO network management model ISO网络管理模型

ISO/OSI model（International Organization for Standardization Open Systems Interconnection model） 国际标准化组织/开放系统互连模型

ISO reference mode ISO参考模式

ISP（Internet Service Provider） 因特网服务提供商

ISR（interactive session routing） 交互会话路由选择

ISR（integrated switch/router） 集成交换/路由器

ISSP（integrated service switching point） 综合业务交换点

ISU（integrated service unit） 综合业务单元

ISUP（ISDN user part） 综合业务数字网用户部分

iTDs（international top level domain names） 国际顶级域名

iTLD（international top level domain names） 国际顶级域名

ITR（Internet talk radio） 因特网无线电对话

ITU（International Telecommunication Union） 国际电信联盟

ITU-TSS（International Telecommunication Union-Telecommunication Standardization Section） 国际电信联盟-电信标准化部门

Iub interface Iub接口，无线网络控制器和移动基站之间的接口

**IVD-LAN**（integrated voice data local area network） 综合声音数据局域网

**IXA**（internet exchange architecture） 互联网交换架构

**IXC**（Internet Exchange Centre） 因特网交换中心

**IXL**（inter exchange link） 局间链路

# J

jabber frame　超时帧
jabbering　无效传送
jabbering procedure　超时过程
JavaBean　JavaBean 软件构件
Java interface definition language (Java IDL)　Java 接口描述语言
Java language　Java 语言
Java management application program interface (JMAPI)　Java 管理应用程序开发接口
Java management extensions (JMX)　Java 管理扩展
Java message service (JMS)　Java 消息传递服务
Java naming & directory interface (JNDI)　Java 命名和目录接口
Java OS　Java 操作系统
Java remote method invocation (Java RMI)　Java 远程方法调用
Java RMI (Java remote method invocation)　Java 远程方法调用
JavaScript language　JavaScript 语言
Java server page (JSP)　Java 服务器页面
Java servlet　Java 服务器小程序
Java virtual machine (JVM)　Java 虚拟机
Java workshop　Java 开发环境

JBIG (Joint Bi-Level Image Coding Experts Group)　联合二值图像编码专家组
Jini　Jini 技术
JMAPI (Java management application program interface)　Java 管理应用程序开发接口
JMS (Java message service)　Java 消息传递服务
JMX (Java management extensions)　Java 管理扩展
JNDI (Java naming & directory interface)　Java 命名和目录接口
job transfer and manipulation (JTM)　作业传送与处理
Joint Bi-lLevel Image Coding Experts Group (JBIG)　联合二值图像编码专家组
joint electronic payments initiative (JEPI)　联合电子支付动议
Joint Photographic Expert Group (JPEG)　联合图像专家小组
JPEG (Joint Photographic Expert Group)　联合图像专家小组
junk mail　垃圾邮件
junk mail filtering　垃圾邮件过滤
junk mail watch　垃圾邮件监测
IUWSN (industry ubiquitous wireless

sensor network) 工业泛在传感网络

JVM (Java virtual machine) Java 虚拟机

JSP (Java server page) Java 服务器页面

# K

**k** 千,1000
**K** 千,1024
**KB（kilobyte）** 千字节
**Kb（kilobit）** 千位
**Kbit（kilobit）** 千位
**KBps（kilobytes per second）** 千字节/秒
**kbps（kilobits per second）** 千位/秒
**KDC（key distribution center）** 密钥分发中心
**key agreement** 密钥协商
**key distribution center（KDC）** 密钥分发中心
**key escrow** 密钥托管
**key escrow agent** 密钥托管代理
**key escrow encryption systems** 密钥托管加密系统
**key pair** 密钥对
**key pair assignment** 密钥对分配
**key pair file** 密钥对文件
**key pair vectors distribution** 密钥对向量分配
**key pair vectors updating** 密钥对向量更新
**KeyPal** 键友
**key routing** 键路由选择
**key table** 键标表
**key-to-address-transformation** 代码地址转换
**keyword stuffing** 关键词堆砌
**KISS（Keep it simple, sir）** "开思",请简单一些
**knowbie** 大虾,网络高手
**Knowbot** 知识机器人
**Knowbot information service（KIS）** 知识机器人信息服务
**knowledge-based artificial neural network** 基于知识的人工神经网络
**knowledge-based fuzzy neural network** 基于知识的模糊神经网络
**knowledge-based neural network** 基于知识的神经网络

# L

label distribution control mode 标记分发控制方式
label distribution protocol (LDP) 标记分发协议
labeled channel 标号信道
label edge router (LER) 标记边缘路由器
labeled multiplexing 带标号的多路复用
labeled statistical channel 带标号的统计性信道
label multiplex 标记复用
label swapping 标识交换
label switch path (LSP) 标签交换路径
label switch router (LSR) 标记交换路由器
label virtual circuit 标记虚拟电路
LACNIC (Lation American and Caribbean Internet Address Registry) 拉丁美洲和加勒比因特网地址注册中心
LACP (link aggregation control protocol) 链路聚合控制协议
lamer 不懂行者,网络新手
LAN (local area network) 局域网
LAN access 局域网接入
LAN access unit (LAU) 局域网接入单元

LAN adapter card 局域网适配卡
LAN administrator 局域网管理员
LAN broadcast 局域网广播
LAN broadcast address 局域网广播地址
LAN description 局域网描述
LANE (local area network emulation) 局域网仿真
LAN emulation (LANE) 局域网仿真
LAN emulation address resolution protocol (LE-ARP) 局域网仿真地址解析协议
LAN emulation client (LEC) 局域网仿真客户机
LAN emulation configuration server 局域网仿真配置服务器
LAN emulation server (LES) 局域网仿真服务器
LAN emulation user network interface (LUNI) 局域网仿真用户-网络接口
LES (LANE server) 局域网仿真服务器
LAN gateway 局域网网关
LAN global address 局域网全局地址
LAN group address 局域网组地址
LAN individual address 局域网个别地址
LAN multicast 局域网组播

LAN multicast address 局域网组播地址
LAN pathname 局域网路径名
LAN recovery 局域网恢复
LAN segment 局域网段
LAN support center 局域网支援中心
LAN switch 局域网交换机
LAP (link access procedure) 链路访问规程
LAP (link access protocol) 链路存取协议
LAP-B (link access procedure-balanced) 平衡链路访问规程
LAP-B frame format 平衡链路访问规程帧格式
LAP-D (link access procedure-D) D信道链路访问规程
LAPM (link access procedure for modems) 调制解调器链路访问规程
large area synchronized CDMA (LAS-CDMA) 大区域同步码分多址
large internet packet (LIP) 网间大数据包
large internetwork packet exchange 大型网间分组交换
large scale point-cloud data 大规模点云数据
large scale network database 大型网络数据库
large scale network database system 大型网络数据库系统
LAS-CDMA (large area synchronized CDMA) 大区域同步码分多址
laser modulation 激光调制
last-in-chain (LIC) 链尾

last mile service 末端服务, 最后1英里服务
Lation American and Caribbean Internet Address Registry (LACNIC) 拉丁美洲和加勒比因特网地址注册中心
lattice-like structure 格状结构
LAU (LAN access unit) 局域网接入单元
lawful interception gateway (LIG) 合法拦截网关
layered system 分层系统
layer entity 层实体
layering model 分层模型
layer management 层管理
layer management entry (LME) 层管理登录项
layer protocol 层协议
layer 1 (L1) 层1, 物理层
layer 2 (L2) 层2, 数据链路层
layer 2 DHCP relay agent 二层DHCP中继代理, 二层动态主机配置协议中继代理
layer 2 forwarding (L2F) 第二层转发协议
llayer 2 switching 第二层交换
llayer 2 tunneling protocol (L2TP) 第二层隧道协议
layer 3 switching 第三层交换
layer 4 switching 第四层交换
LBT (listen before talk) 先听后讲
LCGN (logical channel group number) 逻辑信道组号
LCN (logical channel number) 逻辑通道号
LCN (loosely-coupled network) 松散耦

合网络

LCP (link control protocol) 链路控制协议

LCP (link control procedure) 链路控制规程

LCS (link connection subsystem) 链路连接子系统

LCS (loosely-coupled system) 松散耦合系统

LCSM (link connection subsystem manager) 链路连接子系统管理器

LDAP (lightweight directory access protocol) 轻量目录访问协议

LDC (link data channel) 链路数据信道

LD-CELP (low delay code excited linear prediction) 低时延码激励线性预测（编码）

LDDI (local distributed data interface) 局部分布式数据接口

LDP (label distribution protocol) 标记分发协议

LDT (logical device table) 逻辑设备表

LE (LAN emulation) 局域网仿真

leadership priority 领导权优先级

leaf setup request (LSR) 叶建立请求

leaky bucket algorithm 漏桶算法

LEAP (lightweight extensible authentication protocol) 轻型可扩展认证协议

lean operation 精益运营

learning bridge 学习桥

LE-ARP (LAN emulation address resolution protocol) 局域网仿真地址解析协议

leased line 租用线，专线

least-cost routing 最小成本路由选择

least-loaded path 最小负载通路

least privilege 最小特权

least privilege principle 最小特权原则

least-weight route 最轻路由选择

LEC (local exchange carriers) 本地交换运营商

LECS (local area network emulation configuration server) 局域网仿真配置服务器

LEN (low-entry networking) 低入口连网

LEN connection 低入口连网连接

LEN end node (low-entry networking end node) 低入口连网终端节点

length of a walk 通路长度

LEQ (line equalizer) 线路均衡器

LER (label edge router) 标记边缘路由器

LES (LANE server) 局域网仿真服务器

letter bomb 信件炸弹

leveled fault tolerant routing 容错平衡路由

lexicographic strategy 词典策略

LFAP (lightweight flow accounting protocol) 轻型流量记账协议

LIG (lawful interception gateway) 合法拦截网关

light wave communication 光波通信

lightweight access point protocol (LWAPP) 轻量级接入点协议

lightweight directory access protocol (LDAP) 轻量目录访问协议

lightweight extensible authentication protocol

(LEAP) 轻型可扩展认证协议
lightweight flow accounting protocol (LFAP) 轻型流量记账协议
like 点赞
limited broadcast 有限广播
limited broadcast address 有限广播地址
limited contention protocol 有限竞争协议
limited distance adapter 有限距离适配器
limited-resource link 有限资源链接
limited-resource session 有限资源会话
LIN (local interconnect network) 局部互联网
LIN (logistics information network) 物流信息网络
linear artificial neural network 线性人工神经网络
linear frequency modulation (LFM) 线性调频
linear frequency modulation signal 线性调频信号
linear network 线性网络
linear passive network 线性无源网络
line connection 线路连接
line control 线路控制
line control procedures 线路控制规程
line equalizer (LEQ) 线路均衡器
line grouping 线路分群
line hunting (LH) 寻线
line monitor 线路监听器
line protocol 线路规程
line reflection 传输线反射
line signaling 线路信令
line speed 线路速度
line termination (LT) 线路终端
line termination equipment (LTE) 线路终端装置
line trap 线路陷波器
line-unit-line termination (LULT) 用户-单元-用户终端
line-unit-network termination (LUNT) 用户-单元-网络终端
link access procedure (LAP) 链路访问规程
link access procedure-balanced (LAP-B) 平衡链路访问规程
link access procedure-D (LAP-D) D 信道链路访问规程
link access procedure for modems (LAPM) 调制解调器链路访问规程
link access protocol (LAP) 链路访问协议
link aggregation 链路聚合
link aggregation control protocol (LACP) 链路聚合控制协议
link analysis 链接分析
link budget 链路预算
link connection subsystem (LCS) 链路连接子系统
link connection subsystem manager (LCSM) 链路连接子系统管理器
link control procedure (LCP) 链路控制规程
link control protocol (LCP) 链路控制协议
link data channel (LDC) 链路数据信道
link encryption 链路加密
link exchange 链路交换
link factory 链接工厂
link farm 链接场
link header 链路报头
link integrity verification testing (LIVT)

链路一致性测试
link layer 链路层
link layer discovery protocol (LLDP) 链路层发现协议
link-level addressing 链路级寻址
link load 链路负载
link load balance (LLB) 链路负载均衡
link load balancer 链路负载均衡器
link local unicast address 链路本地单播地址
link loopback 链路环回
link management 链路管理
link management protocol (LMP) 链路管理协议
link margin 链路余量
link protocol (LP) 链路协议
link protocol converter (LPC) 链路协议转换器
link protocol data unit 链路协议数据单位
link resource 链路资源
link resource manager 链路资源管理器
link quality 链路质量
link quality analysis (LQA) 链路质量分析
link-quality aware 铧路质量感知
link quality indicator (LQI) 链路质量指示
link service access point (LSAP) 链路服务访问点
link spamming 垃圾链接,作弊链接
link spectrum efficiency 链路频谱效率
link state (LS) 链路状态
link state advertisement (LSA) 链路状态广播

link state database 链路状态数据库
link state PDU (LSP) 链路状态协议数据单元
link state protocol 链路状态协议
link state routing algorithm 链路状态路由算法
link state routing protocol 链路状态路由协议
link station 链路站
link support layer (LSL) 链路支持层
link test 链路测试
link trailer 链路尾标
LIP (loop initialization protocol) 环路初始化协议
LIP (large internet packet) 网间大数据包
listen before talk (LBT) 先听后讲
listening mode 收听方式
listening silence 收听静寂
listen while talk (LWT) 边听边讲
Litecoin (LTC) 莱特币
LIVT (link integrity verification testing) 链路一致性测试
LLA (logical layer architecture) 逻辑分层体系结构
LLB (link load balance) 链路负载均衡
LLC (logical link control) 逻辑链路控制
LLC protocol (logical link control protocol) 逻辑链路控制协议
LLDP (link layer discovery protocol) 链路层发现协议
LLID (logical link identification) 逻辑链路标识
LLP (low level protocol) 低层协议
LLQ (low latency queuing) 低延时队列
LL2 (link level 2) 链路二级

LMDS (local multipoint distribution service) 本地多点分配业务
LMI (local management interface) 局部管理接口
LMP (link management protocol) 链路管理协议
LMP (loopback mirror protocol) 回送镜像协议
LMTP (local mail transfer protocol) 本地邮件传输协议
load administration 装载监管
load balancer 负载均衡器
load balancing 负载均衡
load balancing mode 负载均衡方式
load sharing computer network 均分负载计算机网络
lobe access unit (LAU) 瓣接入单元
lobe bypass 插接瓣旁路
local area network (LAN) 局域网
local area network (LAN) access 局域网接入
local area network broadcast 局域网广播
local area network emulation (LANE) 局域网仿真
local area network emulation configuration server (LECS) 局域网仿真配置服务器
local area network interconnection 局域网互联
local area network multicast 局域网组播
local area network reference model 局域网参考模型
local area network server 局域网服务器
local area network standardization 局域网标准化
local communication link 本地通信链路
local computer network 局部计算机网络
local directory database 局部目录数据库
local distributed data interface (LDDI) 局部分布式数据接口
local domain name 局部域名
local enforcement point (LEP) 本地增强点
local exchange carriers (LEC) 本地交换运营商
local exchange operation system (LE-OpS) 本地交换运行系统
local host 本地主机
local interconnect network (LIN) 局部互联网
locality of reference 访问局部性
local link 本地链路
local link status information 本地链路状态信息
local-local link 局部与局部链接
locally administered address 局部监管的地址
local mail transfer protocol (LMTP) 本地邮件传输协议
local management interface (LMI) 局部管理接口
local multipoint distribution service (LMDS) 本地多点分配业务
local network 局部网
local newsgroups 局域新闻组
local node 本地节点
local processor link (LPL) 本地处理器连接
local proxy ARP 本地代理 ARP,本地代理地址解析协议

local remote link 本地与远程链接
local routing algorithm 局部路由选择算法
local subnet 本地子网
local subnet scanning 本地子网扫描
local-talk link access protocol (LLAP) 本地聊天链路访问协议
locking protocol 封锁协议
logical carrying network 逻辑承载网络
logical channel 逻辑信道
logical channel group number (LCGN) 逻辑信道组号
logical channel identifier 逻辑通道标识
logical channel number (LCN) 逻辑通道号
logical communication link 逻辑通信链路
logical data link 逻辑数据链路
logical device address (LDA) 逻辑设备地址
logical device table (LDT) 逻辑设备表
logical forward network 逻辑转发网络
logical layer architecture (LLA) 逻辑分层体系结构
logical link 逻辑链路
logical link control (LLC) 逻辑链路控制
logical link control and adaptation layer protocol (L2CAP) 逻辑链路控制和适应层协议
logical link control protocol 逻辑链路控制协议
logical link control sublayer 逻辑链路控制子层
logical link control type 1 (LLC type 1) 1型逻辑链路控制模式
logical link control type 2 (LLC type 2) 2型逻辑链路控制模式
logical link control type 3 (LLC type 3) 3型逻辑链路控制模式
logical link identification (LLID) 逻辑链路标识
logical message 逻辑报文
logical network 逻辑网络
logical network model 逻辑网络模型
logical neural network 逻辑神经网络
logical node 逻辑节点
logical recurrent neural network 逻辑回归神经网络
logical ring 逻辑环
logic ring communication network 逻辑环通信网
logical signaling channel 逻辑信令信道
logical station (LS) 逻辑站
logical topology 逻辑拓扑
logical topology chart 逻辑拓扑图
logical topology reconfiguration 逻辑拓扑重构
logical topology space 逻辑拓扑空间
logical unit (LU) 逻辑单元
logical virtual active network (LVANet) 逻辑虚拟主动网
logic isolation 逻辑隔离
logic multicast 逻辑组播
login domain 登录域
login script 登录脚本
logistics information network (LIN) 物流信息网络
logon server 登录服务器
LOL (laughing out loud) 放声大笑
long end keyword 长尾关键词
longest prefix matching (LPM) 最长前缀

匹配
long-haul network  远程网
longitudinal redundancy check (LRC)  纵向冗余校验
long time evolution (LTE)  长期演进
look-to-buy ration  (网页)查看与购买比
look-to-request ratio  (网页)查看与请求比
loop back address  回送地址
loop back checking  回送校验
loop back mirror protocol (LMP)  回送镜像协议
loop back testing  回送测试
loop computer network  环形计算器网络
loop configuration  环形配置
loop difference signal  环路差值信号
looped dual bus  环接双总线
loop feedback signal  环路反馈信号
loop frame  环路帧
loop gain  环路增益
loop initialization protocol (LIP)  环路初始化协议
loop network  环型网络
loop prevention  环路防止
loop station connector (LSC)  环路站连接器
loop transmission  环路传输
loop transmission frame  环路传输帧
loop wiring concentrator (LWC)  环路连线集中器
loosely-coupled network (LCN)  松散耦合网络
loosely-coupled system (LCS)  松散耦合系统
loss of frame (LOF)  帧丢失

loss of pointer (LOP)  指针丢失
low delay code excited linear prediction (LD-CELP)  低时延码激励线性预测(编码)
low entry networking (LEN)  低入口连网
low entry networking (LEN) end node  低入口连网网络终端节点
low entry networking (LEN) node  低入口连网网络节点
lower layer protocol  低层协议
lower-order path adaptation (LPA)  低阶通道适配
low latency network  低延时网络
low latency queuing (LLQ)  低延时队列
low level control  低层控制
low level protocol (LLP)  低层协议
LP (link protocol)  链路协议
LPA (lower-order path adaptation)  低阶通道适配
LPL (local processor link)  本地处理器连接
LPM (longest prefix match)  最长前缀匹配
LQI (link quality indicator)  链路质量指示
LRC (longitudinal redundancy check)  纵向冗余校验
LS (link status)  链路状态
LS (logical station)  逻辑站
LSA (link state advertisement)  链路状态广播
LSAP (link service access point)  链路服务访问点
LSC (loop station connector)  环路站连接器

LSL (link support layer)　链路支持层
LSP (link state PDU)　链路状态协议数据单元
LSP (label switch path)　标签交换路径
LSR (label switch router)　标记交换路由器
LSR (leaf setup request)　叶建立请求
LT (line termination)　线路终端
LTA (logical transient area)　逻辑暂驻区
LTC (Litecoin)　莱特币
LTE (long time evolution)　长期演进
LTE (line termination equipment)　线路终端装置
LTE-Advanced (LTE-A)　LTE 计划,长期演进计划
LU (logical unit)　逻辑单元
LU connection testing　逻辑单元连接测试
LULT (line-unit-line termination)　用户-单元-用户终端
LU-LU session　LU-LU 会话,逻辑单元之间会话
LU-LU session initiation　LU-LU 会话初启,逻辑单元之间会话初启
LU-LU session termination　LU-LU 会话终止,逻辑单元之间会话终止
LU-LU session type　LU-LU 会话类型,逻辑单元之间会话类型
LUNI (LAN emulation user network interface)　局域网仿真用户-网络接口
LUNT (line-unit-network termination)　用户-单元-网络终端
lurk　潜水
lurker　潜水者
LU service manager　LU 服务管理,逻辑单元服务管理
LVANet (logical virtual active network)　逻辑虚拟主动网
LWAPP (lightweight access point protocol)　轻量级接入点协议
LWC (loop wiring concentrator)　环路连线集中器
LWT (listen while talk)　边听边讲
L1 (layer 1)　层 1,物理层
L2 (layer 2)　层 2,数据链路层
L2CAP (logical link control and adaptation layer protocol)　逻辑链路控制和适应层协议
L2F (layer 2 forwarding)　第二层转发协议
L2TP (layer 2 tunneling protocol)　第二层隧道协议

# M

M (mega)　兆,百万
MA (mobile agent)　移动智能体
MAA (mobile application accelerator)　移动应用加速
MABR (multiplex aggregate bit rate)　多路并合比特率,多路聚合比特率
MAC (media access control)　介质访问控制
MAC (message authentication code)　报文鉴别码
MAC (mandatory access control)　强制访问控制
MAC (multi-access controller)　多路访问控制器
MAC (multiple access communication)　多址通信
MAC address　介质访问控制地址
MAC frame　介质访问控制帧
machine name　机器名字
MAC protocol (media access control protocol)　介质访问控制协议
MAC sublayer (medium access control sublayer)　介质访问控制子层
MAF (management application function)　管理应用功能
mail bomb　邮件炸弹
mail box　邮箱
mailbox facility　邮箱设备

mailbox name　邮箱名
mailbox system　邮箱系统
mail bridge　邮件网桥
mail delivery agent (MDA)　邮件投递代理
mail exchanger (MX)　邮件交换
mail exploder　邮件分发器
mail filtering　邮件过滤
mail folder　邮件夹
mail forwarding (MF)　邮件转发
mail gateway　邮件网关
mail list　邮件列表
mail list server　邮件列表服务器
mail merge　邮件合并
mail notifying server　邮件通知服务器
mail path　邮件路径
mail reader program　邮件阅读程序
mail recall　邮件召回
mail reflector　邮件反射器
mail server　邮件服务器
mail server hosting　邮件服务器托管
mail server rental　邮件服务器租用
mailslot　邮件槽
mail store server　邮件存储服务器
mail transfer agent (MTA)　邮件传输代理
mail transfer protocol　邮件传递协议
mail transmission server　邮件传输服

mail user agent (MUA) 邮件用户代理
mail withdrawn 邮件撤回
main domain name server 主域名服务器
main loop cabling 主环路电缆连接
main network address 主网络地址
main page 主页
main ring path 主环路径
maintenance hook 维护钩
malicious click 恶意点击
malicious software 恶意软件
malicious websites 恶意网站
MAN (metropolitan area network) 城域网
management agent 管理代理
management application function (MAF) 管理应用功能
management application software 管理应用软件
management event notification protocol (MEN) 管理事件指示协议
management information base (MIB) 管理信息库
management service control 管理服务控制
management service provider (MSP) 管理服务提供商
management service 管理服务
managers of managers (MOM) 管理的管理者
mandatory access control (MAC) 强制访问控制
mandatory protection 强制保护
MANET (mobile Ad hoc network) 移动自组织网络
manipulative in the middle attack (MITM) 中间人攻击
manual search 手动搜索
manufacturing automation protocol (MAP) 制造自动化协议
manufacturing automation protocol/technical and office protocol (MAP/TOP) 制造自动化协议/技术与办公协议
manufacturing message service (MMS) 制造报文服务
manufacturing message specification (MMS) 制造报文规范
MAP (manufacturing automation protocol) 制造自动化协议
MAPI (messaging application programming interface) 消息应用程序编程接口
MAP/TOP (manufacturing automation protocol/technical and office protocol) 制造自动化协议/技术与办公协议
marketing blog 营销博客
marketing website 营销网站
marketing without website 无网站营销
Martians "火星人",不速之客分组
martian language 火星文
MASIF (mobile agent system interoperability facility) 移动代理系统互操作环境
MASK (multiple amplitude shift keying) 多进制幅移键控
mass E-mail 群发邮件
master net control station (MNCS) 网内主控站
master node 主节点
master node control 主节点控制

master-slave bus network 主从式总线型网络
master-slave network 主从式网络
master-slave network structure 主从式网络结构
master-slave neural network 主从式神经网络
master-slave synchronization network 主从式同步网
master station (MST) 主站
master virtual circuit (MVC) 主虚拟电路
mathematical markup language (Math ML) 数学标记语言
MATI (multicast access termination identifier) 多点广播接入终端标识
Math ML (mathematical markup language) 数学标记语言
MAU (multistation access unit) 多站访问单元
MAU (monthly activited users) 月活跃用户
MAU (media access unit) 媒体访问单元
MAU (media attachment unit) 媒体连接设备
MaxCR (maximum cell rate) 最大信元速率
maximum allowable packet dropout rate 最大容许数据包丢失率
maximum burst size (MBS) 最大突发长度
maximum cell rate (MaxCR) 最大信元速率
maximum collision domain 最大冲突域
maximum flow 最大流
maximum hop count 最大跳数
maximum segment size (MSS) 最大分段大小
maximum SSCP rerouting count 系统服务控制点最大重定路由计数
maximum stuffing rate 最大填充率
maximum transfer rate 最高传送率
maximum transmission unit (MTU) 最大传输单元
MB (megabyte) 兆字节,百万字节
MBGP (multiprotocol extensions for BGP4) 多协议边界网关协议
Mbit (megabit) 兆位,百万位
MBMS (multimedia broadcast multicast service) 多媒体广播组播业务
MBONE (multicast backbone) 多址传播骨干网
Mbps (million bits per second) 每秒兆位
MBS (maximum burst size) 最大突发长度
Mb/s (million bits per second) 每秒兆位
MBWA (mobile broadband wireless access) 移动宽带无线接入
MC (mobile commerce) 移动商务
MCA (multi-channel access) 多信道选取
MCA (mobile cloud accelerator) 移动云加速器
MCC (mobile cloud computing) 移动云计算
MC-CDMA (multi-carrier code division

multiple access） 多载波码分多址

MCF（Mobile Cloud Forum） 移动云论坛

MC-TDMA（multi-carrier time division multiple access） 多载波时分多址

MCDN（micro cellular data network） 微蜂窝数据网

MCDN（media content distribution network） 媒体内容发布网络

m-commerce 移动电子商务

MCP（multi-stream conversation protocol） 多流对话协议

MCP（media content provider） 媒体内容提供商

MCR（minimum cell rate） 最小信元速率

MCS（multimedia communication system） 多媒体通信系统

MCU（multipoint control unit） 多点控制单元

MDA（mail delivery agent） 邮件投递代理

MDI（medium dependent interface） 媒体相关接口

MDN（mobile directory number） 移动目录号码

mean packet service delay 平均分组服务时延

mean service access delay 平均业务接入迟延

mean service provisioning time 平均业务准备时间

mean service rate 平均服务率

mean service time（MST） 平均服务时间

mean time to service restoral 业务恢复的平均时间

MEC（mobile edge computing） 移动边缘计算

media access control（MAC） 介质访问控制

media access control（MAC）frame 介质访问控制帧

media access control（MAC）protocol 介质访问控制协议

media access control（MAC）sublayer 介质访问控制子层

media access gateway 媒体接入网关

media access unit（MAU） 媒体访问单元

media attachment unit（MAU） 媒体连接设备

media content distribution network（MCDN） 媒体内容发布网络

media content management 媒体内容管理

media content provider（MCP） 媒体内容提供商

media control interface（MCI） 媒体控制接口

media filter 媒体滤波器

media gateway（MG） 媒体网关

media gateway controller（MGC） 媒体网关控制器

media gateway control function（MGCF） 媒体网关控制功能

media gateway control protocol（MGCP） 媒体网关控制协议

media gateway control technology 媒体网关控制技术

media independent interface（MII） 介质无关接口
media network 媒体网络
media network attached storage 媒体网络附属存储
media oriented system transport（MOST） 媒体定向系统传输
media streaming 媒体流，流媒体
media streaming broadcast distribution（MSBD） 流媒体广播发布
media streaming network 流媒体网
media streaming performance 流媒体性能
media streaming server 流媒体服务器
media streaming system 流媒体系统
mediation function 中介功能
mediation function block 中介功能块
medical virtual reality technology 医学虚拟现实技术
medium access constraint 媒体访问约束
medium access control frame 媒体存取控制帧
medium access control layer 媒体访问控制层
medium access control protocol 媒体访问控制协议
medium access delay 媒体访问延迟
medium and small-sized enterprise network 中小型企业网络
medium data rate 中速数据传输率
medium dependent interface（MDI） 媒体相关接口
medium independent interface（MII） 媒体无关接口

medium interface connector（MIC） 媒体接口连接器
MEF（Metro Ethernet Forum） 城域以太网论坛
mega 兆，百万
megabit（Mb） 兆位
megabits per second（Mbps, Mb/s） 每秒兆位
megabyte（MB） 兆字节，百万字节
member firewalls 成员防火墙
member server 成员服务器
MEN（management event notification protocol） 管理事件指示协议
merchant website 商家网站
MERP（mobile enterprise resource planning） 移动企业资源计划
mesh 网络，网格
mesh-isolated bonding network 网状隔离连接网络
mesh network 网状网络
mesh network fault tolerant 网状网络容错
mesh optical network 网状光网
mesh wireless network 网状无线网络
message 报文
message alias 报文别名
message alignment 报文定位
message alignment indicator（MAI） 报文定位标记
message authentication 报文验证
message authentication code（MAC） 报文鉴别码
message block 报文块
message broadcast 消息广播
message broker 报文代理

message buffering synchronization 报文缓冲同步
message buffering synchronization facility 报文缓冲同步机构
message digest 报文摘要
message digest algorithm 报文摘要算法
message domain 报文域
message editor procedure 报文编辑程序工作过程
message element 报文元素
message exhaustion 报文穷举
message feedback 报文反馈
message filtering 报文过滤
message filtering gateway 报文过滤网关
message format 报文格式
message handler (MH) 报文处理程序
message handling 报文处理
message handling protocol 报文处理协议
message handling service (MHS) 报文处理服务
message handling system (MHS) 报文处理系统
message header (MH) 报头
message identifier (MID) 报文标识符
message indicator 报文指示符
message input descriptor (MID) 报文输入描述符
message intercept 报文截取
message intercept processing 报文截取处理
message management program 报文管理程序
message mode 报文方式
message-oriented middleware (MOM) 面向报文的中间件
message-oriented text interchange system (MOTIS) 面向报文的文本交换系统
message passing interface (MPI) 报文传递接口
message pending 报文挂起
message polling 报文轮询
message preparation time 报文准备时间
message priority 报文优先级
message processing time 报文处理时间
message publishing 信息发布
message queue 消息队列
message queue data set 消息队列数据集
message queuing telemetry transport (MQTT) 消息队列遥测传输
message rate 报文速率
message relay 报文中继
message release time 报文发出时间
message response time 报文响应时间
message resynchronization 报文再同步程序
message routing 报文路由选择
message segment 报文段
message sequencing 报文排序
message service 报文业务
message sink 报文宿
message slot 报文槽
message source 报文源
message store (MS) 报文存储

message switching 报文交换
message switching center 报文交换中心
message switching computer access 报文交换计算机访问
message switching concentration (MSC) 报文集中转发
message switching concentration techniques 报文交换集中技术
message switching network 报文交换网络
message switching node 报文交换节点
message switching procedure 报文交换过程
message switch node 报文交换节点
message telecommunication service (MTS) 信息远程通信服务
message text 报文正文
message time objective 客观报文时间
message transfer (MT) 报文传送
message transfer agent (MTA) 报文传送代理
message transfer part (MTP) 报文传送部分
message transfer state 报文传送状态
message transfer system (MTS) 报文传送系统
message transfer time 报文传送时间
message transmitting procedure 报文传输过程
message type (MT) 消息类型
message unit 报文单元
message verification 报文校验
messaging application programming interface (MAPI) 消息应用程序编程接口
messaging service 报文服务
messenger service 信使服务
metanetwork 元网络
metasignaling 元信令
Metro Ethernet Forum (MEF) 城域以太网论坛
metro mirror 城域镜像
metropolitan area network (MAN) 城域网
metropolitan area transport network 城域传送网
metropolitan broadband telecommunication network 城市宽带电信网
metropolitan broadband network 城市宽带网
metropolitan transmission network 城市传输网
MF (mail forwarding) 邮件转发
MFDL (microwave fiber delay line) 微波光纤延迟线
MFMC (multimedia fiber monitor & control) 多媒体光纤监控
MFN (multiple frequency network) 多频网
MFSK (multiple frequency shift keying) 多进制频移键控
MG (media gateway) 媒体网关
MGC (media gateway controller) 媒体网关控制器
MGCP (media gateway control protocol) 媒体网关控制协议
MH (message header) 报头
MH (message handing) 报文处理
MHEG (Multimedia and Hypermedia

Information Encoding Expert Group) 多媒体和超媒体信息编码专家组
MHPW (multi hop pseudo wire) 多跳伪线
MHS (message handling service) 报文处理服务
MHS (message handling system) 报文处理系统
MHz (megahertz) 兆赫(兹)
MIB (management information base) 管理信息库
MIC (medium interface connector) 媒体接口连接器
MIC (middle-in-chain) 链中间单元
micro blog 微博
micro blog marketing 微博营销
micro blog news 微新闻
micro cellular 微蜂窝
micro cellular data network (MCDN) 微蜂窝数据网
micro cellular network 微蜂窝网络
micro cellular propagation model 微蜂窝传播模型
MicroCom networking protocol (MNP) MicroCom 连网协议
microcomputer local area network 微型计算机局部网络
microcomputer master/slave operation 微型计算机主从操作
microcomputer network intelligence 微型计算机网络智能
microcomputer point-of-sale system (MPOSS) 微机销售点系统
micromerchants (网络)微型商人，微商

micromoney (网络)微型货币
micropayment (网络)微型支付款
micropricing (网络)微型定价
microserf 网络奴隶，沉溺于网络的人
Microsite 微型网站
Microsoft media server (MMS) 微软媒体服务器
Microsoft service network (MSN) 微软网络服务
Microsoft transaction server (MTS) 微软事务服务器
micro virtual circuit service 微虚电路服务
microwave network 微波网络
MID (message input descriptor) 报文输入描述符
MID (message identifier) 报文标识符
midamble 训练序列码
middle-in-chain 链中间单元
middle tier 中间层
middleware 中间件
MII (media independent interface) 介质无关接口
.mil 军事机构域名
military information network 军事信息网
military mobile communication network 军事移动通信网
military network (MILNET) 军事网
military optical network 军用光网络
military telecom network 军用电话网
MILNET (military network) 军事网
MIME (multimedia Internet mail extensions) 多媒体因特网邮件扩展
MIME (multipurpose internet mail

extensions) 多用途互联网邮件扩展
MIMO（multiple input multiple output）多进多出
MIN（mobile intelligent network）移动智能网
MIN（mobile identification number）移动标志号码
MIN（multistage interconnection network）多级互联网络
mind your own business（MYOB）不关你的事
minimal netconn switch 极简网络切换
minimum cell rate（MCR）最小信元速率
minimum-cost flow algorithm 最小代价流算法
minimum-cost spanning tree 最小代价生成树
minimum-matching algorithm "最小权匹配"算法
minimum phase frequency shift keying（MSK）最小相位频移键控
minimum shift keying（MSK）最小频移键控
minimum spanning tree（MST）最小生成树
minimum support tree 最小支撑树
minimum weighted path length tree 最小赋权路径长度树
minimum weight routing 最小权重路由选择
Minisite 迷你网站
minor synchronization point service 次同步点服务
mirror server 镜像服务器
mirror site 镜像站点
mirror website 镜像网站
miscellaneous data record（MDR）混杂数据记录
misdelivered bit 误发比特
misdelivered block 误发信息块
missed synchronization 丢失同步信号，漏同步
missed time stamp 丢失时间戳
MITM（manipulative in the middle attack）中间人攻击
mixed-media multilink transmission group 混合媒体多链路传输组
mixed-media system 混合媒体系统
mixed reality（MR）混合现实
mixed reality game 混合现实游戏
mixed reality online game 混合现实网络游戏
mixed reality technology 混合现实技术
mixed routing algorithm 混合路由选择算法
mixed virtual reality technology 混合虚拟现实技术
MLD（multicast listener discovery protocol）组播侦听者发现协议
MLM（multi-level marketing）多层次营销
MLP（multi-link procedure）多链路规程
MLTG（multilink transmission group）多链路传输组
MM（mobile management）移动性管理
MMD（multimedia domain）多媒体域

MMDS (multichannel multipoint distribution service) 多信道多点分配业务
MMG (multimedia management generato) 多媒体信息管理网络自动生成器
MMM (multimedia mail) 多媒体邮件
MMMLTG (mixed-media multilink transmission group) 混合媒体多链路传输组
MMS (Microsoft media server) 微软媒体服务器
MMS (multimedia message service) 多媒体短信服务
MMS (manufacturing message service) 制造报文服务
MMS (manufacturing message specification) 制造报文规范
MNC (mobile network code) 移动网络代码
MNP (MicroCom networking protocol) MicroCom 连网协议
MNP (mobile number portability) 移动号码可携性
MO (mobile originated) 移动发起
mobile Ad hoc network (MANET) 移动自组织网络
mobile agent (MA) 移动智能体
mobile agent server (MAS) 移动代理服务器
mobile agent system interoperability facility (MASIF) 移动代理系统互操作环境
mobile application accelerator (MAA) 移动应用加速
mobile bank 移动银行
mobile-based payment network 手机支付网络, 移动支付网络
mobile broadband wireless access (MBWA) 移动宽带无线接入
mobile business 移动商务
mobile client agent 移动客户代理
mobile cloud 移动云
mobile cloud accelerator (MCA) 移动云加速器
mobile cloud computing (MCC) 移动云计算
Mobile Cloud Forum (MCF) 移动云论坛
mobile collaborative virtual environment (mobile CVE) 移动协作虚拟环境
mobile commerce (MC) 移动商务
mobile communication network 移动通信网
mobile communication satellite 移动通信卫星
mobile computing network 移动计算网络
mobile control agent 移动控制代理
mobile cooperative office 移动协作办公
mobile customer premises network 移动用户驻地网
mobile deployment agent 移动部署代理
mobile digital signature 移动数字签名
mobile E-business 移动电子商务
mobile E-commerce 移动电子商务
mobile E-commerce platform 移动电子商务平台
mobile E-commerce site 移动电子商务网站

mobile edge computing (MEC)　移动边缘计算
mobile emergence terminal　移动应急终端
mobile enterprise resource planning (MERP)　移动企业资源计划
mobile game　移动游戏,手游
mobile identification number (MIN)　移动标志号码
mobile intelligent network (MIN)　移动智能网
mobile intelligent network technology　移动智能网技术
mobile Internet　移动因特网
mobile IP authentication　移动IP认证
mobile IP registration request　移动IP注册申请
mobile management (MM)　移动性管理
mobile network code (MNC)　移动网络代码
mobile network firewall　移动网络防火墙
mobile network signaling　移动网络信令
mobile network terminal　移动网络终端
mobile newspaper　手机报
mobile number portability (MNP)　移动号码可携性
mobile originated (MO)　移动发起
mobile originated short message　移动发起短信息
mobile payment　移动支付
mobile payment agent　移动支付代理
mobile payment network　移动支付网络
mobile payment platform　移动支付平台
mobile payment service　移动支付业务
mobile phone bank　移动电话银行
mobile phone multimedia　移动电话多媒体
mobile phone network　移动电话网络,手机网络
mobile phone network application software　手机网络应用软件
mobile phone payment　手机支付
mobile phone positioning　手机定位
mobile phone user　手机用户
mobile podcasting　移动播客
mobile satellite communication network　移动卫星通信网络
mobile search　移动搜索
mobile secure terminal　移动安全终端
mobile sensor network　移动传感器网络
mobile service switching center　移动业务交换中心
Mobile social network　移动社交网络
mobile station international ISDN number (MSISDN)　移动台国际ISDN号码
mobile streaming media (MSM)　移动流媒体
mobile streaming media-content delivery networks (MSM-CDN)　移动流媒体内容传输网络
mobile streaming media service　移动流媒体业务
mobile switching center (MSC)　移动交换中心
mobile telecommunication service　移动通信业务
mobile terminal (MT)　移动终端
mobile terminal agent　移动终端代理
mobile-to-satellite return channel　移动点至卫星转回信道

mobile ubiquitous computing 移动泛在计算
mobile user agent 移动用户代理
mobile user identifier 移动用户识别
mobile video 移动视频
mobile virtual network operator (MVNO) 移动虚拟网络运营商
mobile virtual private network (MVPN) 移动虚拟专用网
mobile virtual private network operator (MVNO) 移动虚拟专用网络运营商
moblog 移动博客
MOCG (multiplayer online casual game) 大型多人线上游戏
modem (modulator-demodulator) 调制解调器
modem deliminator 调制解调器消除器
modem server 调制解调服务器
moderated mail list 仲裁邮件列表
moderated newsgroup 已仲裁的新闻专题组
moderator 斑竹，论坛版主
modified frequency modulation (MFM) 改进调频制
modulation rate 调制速率
modulation technology 调制技术
modulation types 调制类型
modulator-demodulator (modem) 调制解调器
monthly activited users (MAU) 月活跃用户
MOSPF (multicast open shortest path first) 组播开放最短路径优先

MOST (media oriented system transport) 媒体定向系统传输
MOTIS (message-oriented text interchange system) 面向消息的文本交换系统
mouse trapping 鼠标捕获
Moving Picture Experts Group (MPEG) 活动图像专家组
MPC (multipath channel) 多路径通道
MPEG (Moving Picture Experts Group) 活动图像专家组
MPF (message processing facility) 消息处理机制
.mpg MPEG 格式文件名后缀
MPI (message passing interface) 消息传递接口
MPLS (multiprotocol label switching) 多协议标记交换
MPLS-TP (multiprotocol label switching transport profile) 多协议标记交换传输协议
MPLS VPN technology 多协议标记交换虚拟专用网技术
MPOA (multiprotocol over ATM) ATM 基础上的多协议
MPOSS (microcomputer point-of-sale system) 微机销售点系统
MPR (multiple provider router) 多供应者路由器
MPS (multiple port sharing) 多端口共享
MPSK (multiple phase shift keying) 多值相移键控
MPTN (multi-ptotocol transport network) 多协议传输网络
MQTT (message queuing telemetry transport)

消息队列遥测传输

MR（mixed reality） 混合现实

MRIB（multicast routing information base） 组播路由信息库

MS（multiplex section） 复用段

MSAN（multi-service access network） 综合业务接入网

MSBD（media streaming broadcast distribution） 流媒体广播发布

MSC（mobile switching center） 移动交换中心

MSC（message switching concentration） 报文集中转发

MSDP（multicast source discovery protocol） 组播源发现协议

MSDSL（multirate single pair DSL） 多速率单线对数字用户线路

MSI（multi site interference） 多址干扰

MSISDN（mobile station international ISDN number） 移动台国际ISDN号码

MSK（minimum phase frequency shift keying） 最小相位频移键控

MSK（minimum shift keying） 最小频移键控

MSM（mobile streaming media） 移动流媒体

MSM-CDN（mobile streaming media-content delivery networks） 移动流媒体内容传输网络

MSN（Microsoft service Network） 微软网络服务

MSNF（multisystem networking facility） 多系统网络设施

MSP（multi-subscriber profile） 多用户线路

MSP（management service provider） 管理服务提供商

MSR（multi-service router） 多业务路由器

MSRN（mobile station roaming number） 移动台漫游号码

MSS（maximum segment size） 最大分段大小

MST（minimum spanning tree） 最小生成树

MST（mean service time） 平均服务时间

MSTP（multi-service transport platform） 多业务传送平台

MT（mobile terminal） 移动终端

MTA（mail transfer agent） 邮件传输代理

MTP（message transfer part） 报文传送部分

MTS（message telecommunication service） 信息远程通信服务

MTS（message transfer system） 消息[报文]传输系统

MTS（Microsoft transaction server） 微软事务服务器

MTU（maximum transmission unit） 最大传输单元

MUA（mail user agent） 邮件用户代理

MUD（multi-user dungeon） 多用户"地牢"

MUD object oriented（MOO） 面向对象的多用户网络游戏

multi-access controller（MAC） 多路访

问控制器
multi-address service 多址业务
multi-agent network 多智能体网络
multi-bottleneck network 多瓶颈网络
multi-bottleneck topology 多瓶颈拓扑
multi-carrier code division multiple access（MC-CDMA） 多载波码分多址
multi-carrier time division multiple access（MC-TDMA） 多载波时分多址
multicast access 多播访问
multicast access termination identifier（MATI） 多点广播接入终端标识
multicast address 组播地址
multicast backbone（MBONE） 多址传播骨干网
multicast constrainted routing 组播约束路由
multicast listener discovery protocol（MLD） 组播侦听者发现协议
multicast MAC address 组播MAC地址
multicast media access control address 组播介质访问控制地址
multicast open shortest path first（MOSPF） 组播开放最短路径优先
multicast packet 组播分组
multicast routing information base（MRIB） 组播路由信息库
multicast routing protocol 组播路由协议
multicast routing security 组播路由安全
multicast routing tree 组播路由树
multicast service routing 组播服务路由
multicast source discovery protocol（MSDP） 组播源发现协议
multicast transmission 组播传输
multi-channel access（MCA） 多信道选取
multi-channel multipoint distribution service（MMDS） 多信道多点分配业务
multi-channel routing 多信道路由
multi-channel single-fiber cable 多信道单纤光缆
multi-class network 多级网络
multi-destination network 多目标网络
n.ultidrop 多点
multidrop communication network 多站通信网络
multidrop network 多点网络
multidrop topology 多点拓扑
multi-endpoint connection 多端点连接
multiframe 复帧
multiframe alignment signal 复帧定位信号
multi-functional network database 多功能网络数据库
multi hop network 多跳网络
multi-homed host 多宿主机
multi-homed network 多宿主网络
multi hop pseudo wire（MHPW） 多跳伪线
multi-layer artificial network 多层人工网络
multi-layer control network 多层控制网络

multi-layer logistic network 多层物流网络
multi-layer network 多层网络
multi-layer network architecture 多层网络结构
multi-layer network management 多层网络管理
multi-layer network security 多层网络安全
multi-layer network survivability 多层网络生存性
multi-layer neural network 多层神经网络
multi-layer overlay network 多层覆盖网络
multilevel computer network 多级计算机网络
multi-level marketing (MLM) 多层次营销
multi-level marketing network 多层次营销网络
multi-level network 多级网络
multi-level network marketing 多级网络经营
multi-level neural network 多级神经元网络
multi-level security policy 多级安全策略
multilink 多链路
multilink network 多链路网络
multilink operation 多链路操作
multilink PPP (MLPPP) 多链路点对点协议
multilink procedure (MLP) 多链路规程

multilink transmission group (MLTG) 多链路传输组
multimedia broadband 多媒体宽带
multimedia broadband access network 多媒体宽带接入网
multimedia broadband communication network 多媒体宽带通信网络
multimedia broadcast multicast service (MBMS) 多媒体广播组播业务
multimedia broadband network 多媒体宽带网
multimedia campus network 多媒体校园网络
multimedia communication system (MCS) 多媒体通信系统
Multimedia courseware 多媒体课件
multimedia domain (MMD) 多媒体域
Multimedia Hypermedia Information Encoding Expert Group (MHEG) 多媒体和超媒体信息编码专家组
multimedia instruction network 多媒体教学网络
multimedia interaction network school 多媒体互动网络学校
multimedia Internet mail extensions (MIME) 多媒体因特网邮件扩展
multimedia mail (MMM) 多媒体邮件
multimedia management generato (MMG) 多媒体信息管理网络自动生成器
multimedia message gateway 多媒体信息网关
multimedia message service (MMS) 多媒体短信服务
multimedia fiber monitor & control

（MFMC） 多媒体光纤监控
multimedia network 多媒体网络
multimedia network application 多媒体网络应用
multimedia network equipment 多媒体网络设备
multimedia network model 多媒体网络模型
multimedia optical fibre network 多媒体光纤网络
multimedia optical fibre industrial control network 多媒体光纤工业控制网
multimedia optical fibre 多媒体光纤
multimedia sensor network 多媒体传感器网络
multimedia software 多媒体软件
multimedia streaming server 多媒体流服务器
multimedia teaching network 多媒体教学网
multimedia video streaming 多媒体视频流
multi-mode layer network 多模式分层网络
multi-mode network 多模式网络
multi-network 多重网络
multi-network fusion 多网络融合
mutil-network protocol 多网络协议
multipath channel (MPC) 多路径通道
multipath I/O (MPIO) 多通道 I/O
multipath transmission 多径传输
multiplayer 多人游戏
multiplayer online casual game (MOCG) 大型多人线上游戏

multiplayer online game 多人网络游戏
multiple address 多级地址,多址
multiple address protocol 多址协议
multiple access 多路访问,多路存取
multiple access communication (MAC) 多址通信
multiple access protocols 多点接入协议
multiple access network 多点接入网络
multiple access techniques 多址接入技术
multiple bottleneck network 多瓶颈网络
multiple channel access 多通道存取
multiple constrained QoS multicast routing 多约束 QoS 多播路由
multiple constrained routing selection 多约束路由选择
multiple domain network 多域网络
multiple explicit routers 多显式路由器
multiple frequency network (MFN) 多频网
multiple frequency shift keying (MFSK) 多进制频移键控
multiple gateway 多网关
multiple gateway load balancing 多网关负载平衡
multiple group address 多组地址
multiple hiberarchy neural network 多重神经网络
multiple input multiple output (MIMO) 多进多出
multiple module access 多重模块访问

multiple multicast routing　多组播路由
multiple network　多重网络
multiple networks fusion　多网络融合
multiple neural network　多级神经网络
multiple neural network structure　多神经网络结构
multiple output network　多输出网络
multiple path routing　多路径路由
multiple path routing tree　多路径路由树
multiple phase shift keying (MPSK)　多值相移键控
multiple polling　多重轮询
multiple port network　多端口网络
multiple port sharing (MPS)　多端口共享
multiple protocol communication　多协议通信
multiple protocol over ATM (MPOA)　ATM 上的多协议传输
multiple provider router (MPR)　多供应者路由器
multiple QoS routing　多 QoS 路由
multiple reuse pattern (MRP)　多重复用模式
multiple routing　多路由选择
multiple sequential access method　多顺序存取方法
multiple shared multicast tree　多共享组播树
multiple source multicast　多源组播
multiple subnet routing　多重子网路由
multiple terminal access (MTA)　多终端存取
multiple transmission medium token-ring　多传输介质令牌环
multiple user access　多用户访问
Multiple wavelet network　多重小波网络
multiple wireless router　多路无线路由
multiplex aggregate bit rate (MABR)　多路并合比特率,多路聚合比特率
multiplexer-demultiplexer　复用-分路器
multiplexing device (MUX)　多路复用器
multiplicative decrease　加速递减
multipoint access　多点访问
multipoint configuration　多点配置
multipoint connection　多点连接
multipoint control unit (MCU)　多点控制单元
multipoint network　多点网络
multipoint network topology　多点网络拓扑
multipoint network control system　多点网络控制系统
multipoint-to-multipoint connection　多点到多点连接
multipoint-to-point connection　多点到点连接
multiprotocol communication chip　多协议通信芯片
multiprotocol environment　多协议环境
multiprotocol extension　多协议扩展
multiprotocol extensions for BGP4 (MBGP)　多协议边界网关协议
multiprotocol label switching (MPLS)　多协议标记交换
multiprotocol label switching transport

profile (MPLS-TP) 多协议标记交换传输协议

multiprotocol network 多协议网络

multiprotocol network library 多协议网络库

multiptotocol transport network (MPTN) 多协议传输网络

multiprotocol over ATM (MPOA) ATM基础上的多协议

multiprotocol routing 多协议路由选择

multipurpose internet mail extensions (MIME) 多用途互联网邮件扩展

multipurpose Internet mail extensions type 多用途互联网邮件扩展类型

multirate single pair DSL (MSDSL) 多速率单线对数字用户线路

multi-satellite network 多卫星网络

multi-server network 多服务器网络

multi-service access network (MSAN) 综合业务接入网

multi-service network 多业务网络

multi-service router (MSR) 多业务路由器

multi-service switch 多业务交换机

multi-service transport platform (MSTP) 多业务传送平台

multistage interconnection network (MIN) 多级互联网络

multistation 多站

multistation access unit (MAU) 多站访问单元

multi-stream conversation protocol (MCP) 多流对话协议

multi-subscriber profile (MSP) 多用户线路

multisystem network 多系统网络

multisystem networking facility (MSNF) 多系统网络设施

multithreaded server (MTS) 多线程服务器

multi-net firewall 多网络防火墙

multi-user dungeon (MUD) 多用户"地牢"游戏

multi-user simulated environment (MUSE) 多用户模拟环境

multi vendor integration protocol (MVIP) 多供应商集成协议

multi vendor network 多供应商网络

multiwavelet neural network 多小波神经网络

mutually synchronized network 互同步网络

MUX (multiplexing device) 多路复用器

MVC (master virtual circuit) 主虚拟电路

MVIP (multi vendor integration protocol) 多供应商集成协议

MVNO (mobile virtual network operator) 移动虚拟网络运营商

MVPN (mobile virtual private network) 移动虚拟专用网

MVPNO (mobile virtual private network operator) 移动虚拟专用网络运营商

MYOB (mind your own business) 不关你的事

MX (mail exchanger) 邮件交换

MZAP (multicast-scope zone announcement protocol) 组播区域范围公告协议

# N

NA（neighbor advertisement） 邻居通告
NAC（network access control） 网络访问控制
NAD（network access delay） 网络访问延迟
NADN（nearest active downstream neighbor） 最近活动下游站
NAH（network agent hub） 网络代理集线器
NAK（negative acknowledgement） 否定应答
NAL（network access layer） 网络访问层
name bind protocol（NBP） 名字绑定协议
name caching 名字暂存
name management protocol（NMP） 名字管理协议
name server（NS） 名字服务器
name space 命名空间
name translation 名字翻译
naming context 命名上下文
naming context tree 命名上下文树
NAMPS（narrow band analog mobile phone service） 窄带模拟移动电话服务
NAP（network access point） 网络接入点
NAP（network access protection） 网络访问防御
NAP（network access pricing） 网络访问定价
NAP（network anonymous privacy） 网络匿名隐私权
NAP（national attachment point） （美国）国家附属交换点
NAP（national access point） 国家接入点
NAPT（network address port translation） 网络地址端口转换
NAPT-PT（network address port translation-protocol translation） 网络地址端口转换/协议转换
narrow band（NB） 窄带
narrow band analog mobile phone service（NAMPS） 窄带模拟移动电话服务
narrow band channel 窄带信道，窄带通道
narrow-band code division multiple access（NCDMA） 窄带码分多址
narrow band integrated service digital network（N-ISDN） 窄带综合业务数字网
narrow band noise 窄带噪声
narrow band signal 窄带信号
narrowcast 窄播
narrowcast network 窄播网络
NAS（network access server） 网络接入服务器

NAS (network attached storage) 网络附加存储

NAS (network application support) 网络应用支持

NAS (network application software) 网络应用软件

NASI (NetWare asynchronous service interface) NetWare 异步服务接口

NASREQ (network access server requirements) 网络接入服务器要求

NASS (network attachment sub-system) 网络附加子系统

NAT (network address translation) 网络地址转换

NAT-PT (network address translation-protocol translation) 网络地址转换/协议转换

national access point (NAP) 国家接入点

National Attachment Point (NAP) （美国）国家附属交换点

National Bureau of Standards (NBS) （美国）国家标准局

National Computer Network Emergency Response Technical Team/Coordination Center of China (CNCERT/CC) 中国国家计算机网络应急技术小组/处理协调中心

National Computer Security Center (NCSC) （美国）国家计算机安全中心

national computer system security and privacy advisory board （美国）国家计算机系统安全保密咨询委员会

National Computing and Networking Facility of China (NCFC) 中国国家计算机与网络设施工程

National Cyber Security Alliance (NCSA) 国家网络安全协会

National Information Infrastructure (NII) （美国）国家信息基础设施

National Institute of Standards and Technology (NIST) （美国）国家标准和技术研究所

National Transparent Optical Network Consortium (NTONC) （美国）国家透明光网络联盟

national Internet backbone 国家因特网主干网

national ISDN 国家综合业务数字网

national mobile service center (NMSC) 全国移动业务中心

National Research and Education Network (NREN) （美国）国家研究和教育网

national security information 国家安全信息

national terminal number (NTN) 国家终端号

national top level domain names (nTLDs) 国家顶级域名

national vulnerability database （美国）国家(信息安全)漏洞库

native customer interface architecture 本地客户接口结构

native file format 本机文件格式,原文件格式

native image 本机映像

native instruction 本机指令

native language software 本机语言软件

native mail 本地电子邮件
native mode 本机方式
native multicast 本地组播
native network 本机网络
native object 本机对象
native service 本机服务
native signal processing 本机信号处理,本地信号处理
native signal processor 本机信号处理器
native XML database (NXD) 原生XML数据库
NAU (network addressable unit) 网络可寻址单元
NAU (network accessable unit) 网络可访问单元
NAU (network access unit) 网络存取部件
NAUN (nearest active upstream neighbor) 最近活动上游站
navigate 漫游
navigation bar 导航栏
NA (network address) 网络地址
NAM (network assets management) 网络资产维护
NAT (network assets tracking) 网络资产跟踪
natural information process 自然信息处理
natural language computer 自然语言计算机
natural language database 自然语言数据库
natural language processing 自然语言处理

NBAD (network based anomaly detection) 基于网络的异常检测
NB (narrow band) 窄(频)带
NBI (network binding interface) 网络关联接口
NBMA (non-broadcast multiple access) 非广播多路访问
NBMA address resolution protocol 非广播多路访问地址解析协议
NBP (name bind protocol) 名字绑定协议
NBS (National Bureau of Standards) (美国)国家标准局
NC (network computer) 网络计算机
NCC (network channel conflict) 网络通道冲突
NCCF (network communication control facility) 网络通信控制机制
NCDMA (narrow-band code division multiple access) 窄带码分多址
NCFC (National Computing and Networking Facility of China) 中国国家计算机与网络设施工程
NCK (network computing kernel) 网络计算内核
NCM (network configuration management) 网络配置管理
NCM (network control message) 网络控制消息
n-connection multiplexing n连接多重复用
NCP (network component producer) 网络部件生成器
NCP (network control program) 网络控制程序

NCP (NetWare core protocol) NetWare 核心协议

NCP (network control protocol) 网络控制协议

NCP connectionless SNA transport (NCST) NCP 无连接 SNA 传输,网络控制程序无连接系统网络体系结构传输

NCP packet switching interface 网络控制程序分组交换接口

NCP/EP definition facility 网络控制程序/仿真程序定义设施

NCP/token ring interconnection 网络控制协议/权标环网互联

NCRef (network computer reference profile) 网络计算机参考概要

NCRP (network computer reference profile) 网络计算机参考配置

NCS (network communication server) 网络通信服务器

NCS (network computing system) 网络计算系统

NCS (network control system) 网络控制系统

NCS (network call signaling) 网络呼叫信令

NCSA (National Cyber Security Alliance) 国家网络安全协会

NCSC (National Computer Security Center) (美国)国家计算机安全中心

NCSI (network communication service interface) 网络通信服务接口

NCST (NCP connectionless SNA transport) NCP 无连接 SNA 传输

NCVM (networked collaborative virtual manufacturing) 网络化协作虚拟制造

NCW (network center war) 网络中心战

ND (neighbor discovery) 邻居发现

NDF (network dynamic function) 网络动态功能

NDIS (network driver interface specification) 网络驱动器接口规范

NDM (network data mover) 网络数据移动器

NDMP (network data management protocol) 网络数据管理协议

NDP (neighbor discovery protocol) 邻居发现协议

NDP (Nortel discovery protocol) 北电发现协议

NDP (network diagnostic process) 网络诊断过程

NDR (network data reduction) 网络数据简化

NDR (network data representation) 网络数据表示

NDS (Novell directory service) Novell 目录服务

NDS (NetWare directory services) NetWare 目录服务

NDS (network data series) 网络数据系列

NDS (network domain security) 网络域安全

NDT (network data throughput) 网络数据吞吐量

NE (network element) 网络元素

near-end cross talk (NEXT) 近端串扰

nearest active downstream neighbor

（NADN） 最近活动下游站
nearest active upstream neighbor (NAUN) 最近活动上游站
nearest neighbor algorithm (NN) 最邻近算法
near field communication (NFC) 近场通信
near neighbor mesh network 近邻域网格形网络
NEC (networked embedded computing) 联网嵌入式计算
necessary bandwidth 必要带宽
necroposting 灌水
NEF (network element function) 网元功能
negotiable BIND 可商谈 BIND，可商谈伯克利因特网名字域
negotiable link station 可商谈链接站
negative acknowledgement (NAK) 否定应答
negative response 否定响应
negotiation 协商
negotiation algorithm 协商算法
negotiation metadata 协商元数据
neighbor acquisition 相邻搜索
neighbor advertisement (NA) 邻居通告
neighbor cache table 邻居缓存表
neighbor-connectivity 相邻连通性
neighbor discovery (ND) 邻居发现
neighbor discovery protocol (NDP) 邻居发现协议
neighbor domain of node 节点邻域
neighbor exchange protocol 邻居交换协议

neighbor gateway 邻近网关
neighborhood 邻域
neighborhood averaging 邻域平均法
neighborhood information frame 邻域信息帧
neighborhood logic 邻域逻辑
neighboring router 相邻路由器
neighbor node 邻近节点
neighbor node discovery 邻节点发现
neighbor node polling 邻节点轮询
neighbor notification 邻居通知
neighbor node selection 邻居节点选择
neighbor node table 邻节点列表
neighbor notification 邻站通知
neighbor quequl 邻居队列
neighbor set 邻区组
neighbor solicitation (NS) 邻居请求
neighbourhood pattern sensitive fault 邻近模式敏感故障
NEL (network element layer) 网络元素层
NEMO (network mobility) 网络移动性
nested sparse grid 嵌套式稀疏网格
.net 网络机构域名
.NET .NET体系结构
net abuse 网络滥用
NetBEUI (NetBIOS extended user interface) 网络基本输入/输出系统扩充用户接口
NetBEUI transport NetBEUI 传输，网络基本输入/输出系统扩充用户接口传输
NetBIOS (network basic I/O system) 网络基本输入/输出系统

NetBIOS extended user interface (NetBEUI)  网络基本输入/输出系统扩充用户接口
NetBIOS frame control protocol  网络基本输入输出系统帧控制协议
NetBIOS frame protocol  网络基本输入输出系统帧协议
NetBIOS framing control protocol  网络基本输入输出系统成帧控制协议
NetBIOS name server  网络基本输入输出系统名字服务器，NetBIOS 名字服务器
NetBIOS over TCP/IP  在 TCP/IP 上的网络基本输入/输出系统
NetBIOS protocol  网络基本输入输出系统协议
NETBLT (network block transfer)  网络数据块传送
NetBook  上网本
net boot  网络启动
net boundary  网边界
net broadcast service  网络广播服务
net buffer  网络缓冲器
net business architecture  网络商务体系结构
netcast  网播
net celebrity  网络名人
net citizen (netizen)  网民
net computing  网络计算
net computing time  网络计算时间
net control communication subsystem  网络控制通信子系统
net control group  网络控制组
net control program  网络控制程序
net control station  网络控制站
net control terminal  网络控制终端
net coordination message  网络协调报文
net coordination station message  网络协调站报文
net cruiser  网络漫游器
NetCut  网络剪刀手
NetDDE (network dynamic data exchange)  网络动态数据交换
netdead  网上死亡
net difference report  网络差异报告
net effect  网络效益
.NET enterprise server  .NET 企业服务器
net etiquette (netiquette)  网络礼仪
net file system  网络文件系统
netfilter  网络过滤器
.NET framework  .NET 框架
NETGEN (network generation)  网络生成
net god  网神
net head  网络迷
net heavy  网络能手
NETID (network identifier)  网络标识
netiquette (net etiquette)  网络礼仪
netiquette guidelines  网络礼仪指南
netizen (net citizen)  网民
netkey  网络密钥
net kook  网上怪人
netlogon service  网络登录服务
net monitor  网络监控程序
net neutrality  网络中立
net neutrality rules  网络中立规则
netnews  网络新闻组
NetOLTP (net online transaction processing)  网络联机事务处理

NetPC  网络个人计算机
net personality  网络名人
net police  网络警察
netpopup service  网络弹出式服务
Netscape  网景公司,网景程序
Netscape application programming interface  网景应用程序编程接口
Netscape application server  网景应用服务器
Netscape Browser  网景浏览器
Netscape Communication Corporation  网景通信公司
Netscape Navigator  网景网络导航
Netscape ONE(Netscape open network environment)  网景开放式网络环境
Netscape open network environment(Netscape ONE)  网景开放式网络环境
Netscape server application programming interface(NSAPI)  网景服务器应用编程接口
net session  网络会话
net simulation language  网络模拟语言
netspeak  网上用语,网络用语
net surfing  网络冲浪
net synthesis  网络综合症
net theory  网络理论
net top box  网络机顶盒
net transmit  网络发送器
net violent game  网络暴力游戏
NetWare access server  NetWare 访问服务器
NetWare asynchronous service interface(NASI)  NetWare 异步服务接口
NetWare core protocol(NCP)  NetWare 核心协议
NetWare directory database(NDD)  NetWare 目录数据库
NetWare directory service(NDS)  NetWare 目录服务
NetWare link service protocol(NLSP)  NetWare 链路服务协议
NetWare loadable module  NetWare 可装载模块
NetWare management system(NMS)  NetWare 管理系统
Netware multiprotocol router  NetWare 多协议路由器
NetWare network operating system(NOS)  NetWare 网络操作系统
NetWare users international(NUI)  NetWare 国际用户组织
network(NET)  网络
network abort  网络异常中止
network accessable unit(NAU)  网络可访问单元
network access configuration function  网络接入配置功能
network access control(NAC)  网络访问控制
network access controller  网络接入控制器
network access control strategy  网络访问控制策略
network access delay(NAD)  网络访问延迟
network access device  网络接入设备
network access facility  网络接入设施
network access flow control  网络接入流量控制

network access gateway 网络接入网关
network accessibility 网络接入能力
network access identifier 网络接入标识符
network accessing interface protocol 进网接口协议,网络接入接口协议
network access layer (NAL) 网络访问层
network access machine (NAM) 网络存取机
network access node 网络访问节点
network access permission 网络访问权限
network access point (NAP) 网络接入点
network access pricing (NAP) 网络访问定价
network access protection (NAP) 网络访问防御
network access protocol 网络访问协议
network access server (NAS) 网络接入服务器
network access server requirements (NASREQ) 网络接入服务器要求
network access unit (NAU) 网络存取部件
network accounting 网络记账
network adaptation layer 网络适应层
network adapter 网络适配器
network adapter board 网络适配板
network addiction 网络成瘾
network address (NA) 网络地址
network addressable unit (NAU) 网络可寻址单元
network addressing extension 网络寻址扩展
network address port translation (NAPT) 网络地址端口转换
network address port translation-protocol translation (NAPT-PT) 网络地址端口转换/协议转换
network address translation (NAT) 网络地址转换
network address translation-protocol translation (NAT-PT) 网络地址转换/协议转换
network administmtion computer center 网络管理计算机中心
network administration implementation program 网络管理应用程序
network administration storage facility 网络管理存储设施
network administrator 网络管理员
network advertisement alliance 网络广告联盟
network advertising 网络广告
network advertising agency 网络广告代理
network advertising media 网络广告媒体
network agent 网络代理
network agent card 网络代理卡
network agent hub (NAH) 网络代理集线器
network alarm 网络告警,网络报警
network algorithm 网络算法
network analog 网络模拟
network analysis 网络分析
network analysis control surveillance 网络分析控制监督
network analysis for system application 系统应用网络分析

network analysis model 网络分析模型
network analysis program 网络分析程序
network analysis theory 网络分析理论
network analysis unit 网络分析设备
network analyzer 网络分析器
network anchor 网络主播
network and operations plan 网络和运行计划
network and switching subsystem 网络和交换子系统
network and switching subsystem (NSS) 网络交换子系统
network and system management process 网络和系统管理过程
network and systems support 网络与系统支持
network and trunk control 网络与干线控制
network anomaly traffic detection 网络异常流量检测
network anonymizer 网络匿名
network anonymous privacy (NAP) 网络匿名隐私权
network anonymous protocol 网络匿名协议
network anonymous sources 网络匿名信源
network anonymous speech 网络匿名言论
network answer 网络应答
network answer question 网络答疑
network answer question system 网络答疑系统
network anti virus 网络反病毒

network application 网络应用
network application function (NAF) 网络应用功能
network application platform 网络应用平台
network application server 网络应用服务器
network application software (NAS) 网络应用软件
network application support (NAS) 网络应用支持
network appointment 网络预约
network architecture 网络体系结构
network architecture functional model 网络体系结构功能模型
network architecture design 网络体系结构设计
network assets management (NAM) 网络资产维护
network assets tracking (NAT) 网络资产跟踪
network assisted GPS 网络辅助全球定位系统
network asynchronous communication 网络异步传播
network attached storage (NAS) 网络附加存储
network attached storage server 网络附加存储服务器
network attachment control functions 网络附加控制功能
network attachment sub-system (NASS) 网络附加子系统
network attack 网络攻击
network attack architecture 网络攻击

体系

network attack behaviors analysis 网络攻击行为分析

network attack classification 网络攻击分类

network attack prevention 网络攻击防御

network attack technology 网络攻击技术

network audio visual 网络视听

network authentication 网络验证

network average delay time 网络平均延时

network awareness 网络意识,网络识别,网络感知

network backbone 主干网,网络主干

network backup 网络备份

network backup system 网络备份系统

network banner advertising 网路标题广告

network bandwidth 网络带宽

network bandwidth capacity 网络带宽容量

network bandwidth detection 网络带宽检测

network bandwidth measurement 网络带宽测量

network bandwidth throttling 网络带宽限制

network bank 网络银行

network-based anomaly detection 基于网络的异常检测

network-based intrusion detection 基于网络的入侵检测

network-based virtual private network 基于网络的虚拟专用网

network basic I/O system (NetBIOS) 网络基本输入/输出系统

network behavior 网络行为

network behavioral norms 网络行为规范

network behavior analysis 网络行为分析

network behavior control system 网络行为控制系统

network behavior feature 网络行为特点

network behavior monitoring 网络行为监控

network behind the server 服务器背后的网络

network bibliography 网络目录学

network binding 网络绑定,网络联编

network binding interface (NBI) 网络关联接口

network blocking probability 网络阻塞率

network block transfer (NETBLT) 网络数据块传送

network block transfer protocol 网络数据块传送协议

network border gateway 网络边界网关

network border node 网络边界节点

network border security 网络边界安全

network bottleneck 网络瓶颈

network bridge 桥接器,网桥

network broadcast 网络广播

network broadcast address 网络广播地址

network broadcast protocol 网络广播协议
network buffer 网络缓冲器
network bus controller 网络总线控制器
network bus system 网络总线系统
network busy hour 网络忙时
network byte order 网络字节顺序
network byte ordering 网络字节排序
network cache 网络缓存
network call 网络呼叫
network call center 网络呼叫中心
network call processing subsystem 网络呼叫处理子系统
network call signaling (NCS) 网络呼叫信令
network capability analysis 网络性能分析
network capacity 网络容量
network capacity expansion 网络容量扩充
network capacity planning 网络容量规划
network capacity utilization ratio 网络容量利用率
network card 网卡
network casual game 网络休闲游戏
network CD server 网络光盘镜像服务器
network channel conflict (NCC) 网络通道冲突
network channel interface 网络信道接口
network channel management system 网络信道管理系统

network chart 网络图
network chat 网络聊天
network cheating 网络欺骗
network class 网络类型
network clear indication delay 网络拆线指示延迟
network clock 网络时钟
network closed loop control 网络闭环控制
network closed loop control system 网络闭环控制系统
network coding 网络编码
network cognition 网络认知,网络认识
network color code 网络色码
network common carrier 网络公共载体,网络运营商
network communication 网络传播
network communication agent 网络通信代理
network communication control facility (NCCF) 网络通信控制机制
network communication engine 网络通信引擎
network communication layer 网络通信层
network communication processor 网络通信处理机
network communication server (NCS) 网络通信服务器
network communication service interface (NCSI) 网络通信服务接口
network community 网络社区
network compatibility 网络兼容性
network component producer (NCP)

网络部件生成器

network computer (NC) 网络计算机

network computer interface 网络计算机接口

network computer reference profile (NCRP) 网络计算机参考配置

network computing architecture 网络计算系统结构

network computing framework 网络计算架构

network computing kernel (NCK) 网络计算内核

network computing system (NCS) 网络计算系统

network concurrent server 网络并发服务器

network configuration 网络配置

network configuration management (NCM) 网络配置管理

network configuration parameter 网络配置参数

network congestion 网络拥塞

network congestion charge 网络拥塞费用

network congestion control 网络拥塞控制

network congestion control algorithms 网络拥塞控制算法

network congestion management 网络拥塞管理

network congestion signal 网络拥塞信号

network connection (NC) 网络连接

network connection link 网络连接链路

network connection port 网络连接端口

network connectivity rule 网络连通性规则

network constant 网络常数

network consumption 网络消费

network consumption mentality 网络消费心理

network consumption theory 网络消费理论

network contention 网络争用

network control agent 网络控制代理

network control block (NCB) 网络控制块

network control center 网络控制中心

network control language 网络控制语言

network control message (NCM) 网络控制消息

network control phase 网络控制阶段

network control port 网络控制端口

network control processor 网络控制处理机

network control program (NCP) 网络控制程序

network control program BSC or SS session 网络控制程序二进制同步通信或起停式通话

network control program generation 网络控制程序生成

network control program major node 网络控制程序大节点

network control program node 网络控制程序节点

network control program station 网络

控制程序站

network control protocol (NCP) 网络控制协议

network control server 网络控制服务器

network control signaling rate 网络控制信令速率

network control signaling unit 网络控制信令装置

network control system (NCS) 网络控制系统

network control unit (NCU) 网络控制单元

network core switch 网络核心交换机

network coverage 网络覆盖

network coverage area 网络覆盖范围

network coverage rate 网络覆盖率

network convergence 网络融合

network convergence based on IP 基于IP的网络融合

network coordinating center 网络协调中心

network coordinating message 网络协调报文

network coordination 网络协调

network coprocessor 网络协处理器

network copyright 网络版权,网络著作权侵权

network copyright authentication 网络版权认证

network copyright infringement 网络侵权行为

network copyright infringement evidence 网络侵权证据

network copyright protection 网络版权保护

network copyright tort 网络著作权侵权

network core protocol (NCP) 网络核心协议

network covert channel model 网络隐蔽通道模型

network credit 网络信用

network credit certification 网络信用证

network credit evaluation 网络信用评价

network credit system 网络信用体系

network crowdfunding 网络众筹

network cryptographic device 网络密码装置

network cryptosystem 网络密码体制

network culture consumption 网络文化消费

network data 网络数据

network database 网络数据库

network database design 网络数据库设计

network database management 网络数据库管理

network database search 网络数据库检索

network database system 网络数据库系统

network data dictionary 网络数据字典

network data encryption 网络数据加密

network data link control 网络数据链路控制

network data reduction (NDR) 网络数据简化
network data management protocol (NDMP) 网络数据管理协议
network data mover (NDM) 网络数据移动器
network data packet 网络数据包
network data packet capturing 网络数据包捕获
network data packet dropout 网络数据包丢失
network data processor 网络数据处理器
network data protection 网络数据保护
network data reduction 网络数据简化
network data reporting 网络数据报告
network data representation 网络数据表示
network data series (NDS) 网络数据系列
network data service 网络数据服务
network data structure 网络数据结构
network data throughput (NDT) 网络数据吞吐量
network data translator 网络数据翻译程序
network data transmission 网络数据传输
network data unit 网络数据单元
network deadlock 网络死锁
network definition 网络定义
network definition language 网络定义语言
network definition procedure 网络定义过程
network delay time 网络延迟
network density 网络密度
network-dependent call connection delay 网络相关呼叫连接延迟
network description table 网络描述表
network design 网络设计
network design algorithm 网络设计算法
network design and management system 网络设计和管理系统
network design criteria 网络设计指标
network destination code 网络目的地代码
network detection 网络检测
network detector 网络检测仪
network determined user busy 网络确定用户忙
network development system 网络开发系统
network device driver 网络设备驱动器
network device installation wizard 网络设备安装向导
network diagnosis 网络诊断
network diagnostic control 网络诊断控制
network diagnostic engine 网络诊断引擎
network diagnostic process (NDP) 网络诊断过程
network diffuse 网络扩散
network digit media 网络数字媒体
network dimensioning 网络规划
network directory 网络目录

network directory database 网络目录数据库
network directory service 网络目录服务
network directory system 网络目录系统
network disaster backup 网络灾害备份
network disconnection 网络断线
network disconnection protection 网络断线保护
network distribution equipment 网络分配设备
network distribution system 网络分布系统
network domain 网络域
network domain security (NDS) 网络域安全
network drills 网络演习
network drive 网络驱动器
network driver interface 网络驱动程序接口
network driver interface specification (NDIS) 网络驱动器接口规范
network duality 网络对偶性
network dynamic 网络动态
network dynamic configuration 网络动态配置
network dynamic data exchange (NetDDE) 网络动态数据交换
network dynamic model 网络动态模型
network dynamic reconfiguration 网络动态重组
network dynamics 网络动力学
networked collaborative virtual manufacturing (NCVM) 网络化协作虚拟制造
networked economy 网络经济
network edge 网络边缘
network edge controlling 网络边缘控制
network electronic mail 网络电子邮件
network element (NE) 网络元素
network element function (NEF) 网元功能
network element function block 网络单元功能块
network element layer (NEL) 网络元素层
network encryption 网络加密
network encryption mode 网络加密方式
network end-point 网络端点
network engine 网络引擎
network entity identifier 网络实体标识符
network entry point 网络入口点
network environment 网络环境
network equalizer 网络均衡器
network equivalent analysis 网络等效分析
network error management facility 网络错误管理设施
network etiquette (Netiquette) 网络礼仪
network exchange 网络交换(机)
network expansion option 网络扩充选项
network expert advisory tool 网络专家咨询工具
network extension 网络扩展,网络

扩充

network extension unit (NEU) 网络扩展设备

networked multimedia 网络多媒体

networked organization 网络组织

networked sales 网络销售

networked society 网络社会

networked transaction 网上交易

network electronic mail 网络电子邮件

network element (NE) 网络元素

network element configuration 网元配置

network element function (NEF) 网元功能

network element function block 网络单元功能块

network element layer (NEL) 网络元素层

network element logs 网元登录

network element management system 网络单元管理系统

network element object 网元对象

network element rent 网络元素出租

network element switch 网元交换机

network emoticonal symbol 网络表情符号

network emulation attack 网络仿真攻击

network emulation profile 网络仿真配置文件

network encryption 网络加密

network encryption card 网络加密卡

network encryption mode 网络加密方式

network encryption system 网络加密系统

network end-point 网络端点

network engine 网络引擎

network engineering 网络工程

network engineering design 网络工程设计

network entity identifier 网络实体标识符

network entry point 网络入口点

network environment 网络环境

network environment consciousness 网络环境意识

network equalizer 网络均衡器

network equilibrium analysis 网络平衡分析

network equipment configuration 网络设备配置

network equivalent analysis 网络等效分析

network error management facility 网络错误管理设施

network etiquette (Netiquette) 网络礼仪

network exchange 网络交换(机)

network expansion option 网络扩充选项

network expert advisory tool 网络专家咨询工具

network extension 网络扩展,网络扩充

network extension mode 网络扩展模式

network extension unit (NEU) 网络扩展单元

network failure 网络失效

network fault intelligent diagnosis 网络故障智能诊断
network fault management (NFM) 网络故障管理
network fault management system 网络故障管理系统
network fault tolerant 网络容错
network faxing 网络传真
network file access method 网络文件存取法
network file access protocol 网络文件存取协议
network file server 网络文件服务器
network file system (NFS) 网络文件系统
network file system implementation 网络文件系统实现
network firewall 网络防火墙
network firewall technology 网络防火墙技术
network flow 网络(信息)流
network formation 网络组建,组网
network formation protocol 组网协议
network forming element 网络形成元素
network frame 网络帧
network frequency 网络频率
network front end 网络前端
network function 网络函数,网络功能
network function symbol 网络函数符号
network fusion 网络融合
network game 网络游戏,网游
network game indulgence 网络游戏成瘾

network game player 网络游戏用户
network game server 网络游戏服务器
network gateway 网络网关
network gateway accounting (NGA) 网络网关记账
network general control protocol (NGCP) 网络通用控制协议
network generation (NETGEN) 网络生成
network geometric topology 网络几何拓扑
network graphics protocol 网络图形协议
network group control 网络群控
network guard system 网络守护系统
network handoff 网络切换
network harddisk 网络硬盘,网盘
network hiccup 网络打嗝,网络出错
network hierarchy and multilayer survivability 网络层次和多层生存能力
network hijacker 网络劫持
network hospital registration booking 网络医院预约挂号
network host 网络主机
network human-machine interface subsystem 网络人机接口子系统
network identifier (NETID) 网络标识符
network identity 网络标识
network imaging server 网络映像服务器
network implementation language 网络实现语言
network inbreak 网络入侵

network independence 网络独立性
network in-dialing 网络拨入
network indicator 网络指示器
network information assets 网络信息资产
network information center (NIC) 网络信息中心
network information consulting service (NICS) 网络信息咨询服务
network information encryption 网络信息加密
network information flow volume 网络信息流量
network information integration 网络信息集成
network information management system 网络信息管理系统
network information media 网络信息媒体
network information penetration 网络信息渗透
network information penetration detection technology 网络信息渗透检测技术
network information resource 网络信息资源
network information retrieval 网络信息检索
network information rule 网络信息规范
network information service (NIS) 网络信息服务
network information space 网络信息空间
network information traffic 网络信息流量
network information transmission 网络信息传输
network infrastructure equipment 网络基础设施设备
networking 网络连接,联网
networking capabilities 连网能力
networking convention 连网约定
networking coprocessor 连网协处理器
networking protocol 连网协议
network initialization 网络初始化
network inspecting 网络检测
network install management 网络安装管理
network integration 网络集成
network integrity 网络完整性
network integrity control system 网络完整性控制系统
network intelligence capabilities enhancement (NICE) 智能型网络增强架构
network intelligence engine 网络智能引擎
network intelligent 网络智能
network intelligent memory-based grid computing 基于网络智能内存的网格计算
network interactive voice response 网络交互话音响应
network interconnect 网络互联
network interface 网络接口
network interface card (NIC) 网络接口卡,网卡
network interface configuration 网络端口配置
network interface definition language

(NIDL) 网络接口定义语言
network interface layer 网络接口层
network interface machine (NIM) 网络接口机
network interface module (NIM) 网络接口模块
network interface signaling 网络接口信令
network interface switch 网络接口交换机
network interface unit 网络接口部件
network intrusion 网络入侵
network intrusion abnormal detection 网络入侵异常检测
network intrusion deception 网络入侵诱骗
network intrusion detection model 网络入侵检测模型
network intrusion detection system (NIDS) 网络入侵检测系统
network intrusion immune system 网络入侵免疫系统
network intrusion prevention system (NIPS) 网络入侵预防系统
network inventory 网络资产
network inventory management 网络资产管理
network islanding 网络孤岛
network isolation 网络隔离
network isolation circuit 网络隔离电路
network isolation device 网络隔离装置
network isolation environment 网络隔离环境
network isolation system 网络隔离系统
network isolation technology 网络隔离技术
network job 网络作业
network job entry (NJE) 网络作业输入
network journal 网络期刊
network kernel extension 网络核心扩展(程序)
network key 网络密钥
network killing virus 网络杀毒
network knowledge 网络知识
network knowledge copyright protection 网络知识产权保护
network language 网络语言
network language violence 网络语言暴力
network latency 网络延时
network layer 网络层
network layer address 网络层地址
network layer address management 网络层地址管理
network layer architecture 网络层次结构
network layer gateway 网络层网关
network layer IP protocol 网络层 IP 协议
network layer control protocol 网络层控制协议
network layer protocol (NLP) 网络层协议
network layer protocol identification (NLPI) 网络层协议标识符
network layer protocol phase 网络层

协议阶段
network layer security protocol (NLSP) 网络层安全协议
network layer topology 网络层拓扑
network layer topology discovery 网络层拓扑发现
network layer topology measurement 网络层拓扑测量
network layout planning 网络布局规划
network levels 网络层次
network life cycle 网络生命周期
network life time 网络生存时间
network limit 网络权限
network link capacities 网络链路容量
network link controller 网络链路控制器
network link test 网络链路测试
network link transfer control 网络链路传输控制
network loadable module 网络可装载模块
network load analysis 网络负载分析
network load sharing 网络负载共享
network load test system 网络负荷测试系统
network load topology 网络负载拓扑
network location awareness service 网络位置感知服务
network location identifier 网络位置标识
network lock-up 网络死锁
network log 网络日志
network logical address (NLA) 网络逻辑地址

network logical data manager (NLDM) 网络逻辑数据管理器
network logic topology 网络逻辑拓扑
network logon 网络登录,网络注册
network logon failed 网络登录失败
network loop 网络环路
network low level data 网络底层数据
network low level protocol 网络底层协议
network main page 网络主页
network maintenance strategy 网络维护策略
network management 网络管理
network management agent 网络管理代理
network management boundary 网络管理边界
network management center (NMC) 网络管理中心
network management console 网络管理控制台
network management consulting service (NMCS) 网络管理咨询服务
network management entity 网络管理实体
Network Management Forum (NMF) 网管论坛
network management gateway (NMG) 网络管理网关
network management information gathering device 网管信息采集设备
network management information model (NMIM) 网络管理信息模型
network management interface emulation 网管接口仿真

network management layer (NML) 网络管理层

network management model 网络管理模型

network management-oriented transaction 面向网络管理的事务(处理)

network management protocol (NMP) 网络管理协议

network management server 网络管理服务器

network management station (NMS) 网络管理站

network management system (NMS) 网络管理系统

network management system architecture 网络管理系统体系结构

network management tool 网络管理工具

network management vector transport (NMVT) 网络管理向量传输

network management Web server 网络管理 Web 服务器

network manager 网络管理器

network matching 网络匹配

network maximum segment size 网络最大段长度

network maximum service volume 网络最大服务量

network marketing 网络营销

network marketing and promotion 网络营销与推广

network marketing outlet 网络营销渠道

network marketing service 网络营销服务

network measurement 网络测量

network media 网络媒体

network media violence 网络媒体暴力

network meltdown 网络瘫痪

network mix 网络组合,网路混合

network mobility 网络移动性

network mobility agent 网络移动代理

network mode 网络方式

network model 网状模型

network modem 网络调制解调器

network monitor 网络监控器

network monitoring public opinion 网络舆情监测

network moral 网络道德

network multimedia 网络多媒体

network multimedia application 网络多媒体应用

network multimedia application model 网络多媒体应用模型

network multimedia courseware 网络多媒体课件

network multimedia player 网络多媒体播放器

network multipoint connection 网络多点连接

network name 网络名

network name server 网络名字服务器

network neighborhood 网络邻域

Network new media 网络新媒体

network news 网络新闻组

network news transfer protocol (NNTP) 网络新闻传递协议

network node 网络节点

network node activity 网络节点活跃度

network node connection 网络节点连接
network node domain 网络节点域
network node interface (NNI) 网络节点接口
network node server (NNS) 网络节点服务器
network noise 网络噪声
network number 网络号
network number exchange 网络编号交换(码)
network object 网络对象
network objective 网络目的
network of resource information 资源信息网络
Network-on-Chip (NoC) 片上网络
network online 网络在线
network online classroom 网络在线课堂
network online consultation 网络在线咨询
network online examining 网络在线检查
network online friends 网络在线朋友
network online payment 网络在线支付
network online test 网络在线考试
network operating procedure 网络操作过程
network operating system (NOS) 网络操作系统
network operating system type 网络操作系统类型
network operation center (NOC) 网络运行中心
network operation support system (NOSS) 网络运营支撑系统
network operator (NO) 网络运营商
network operator command 网络操作员命令
network operator console 网络操作员控制台
network operator logon 网络操作员注册
network operator service 网络操作员服务
network optimization 网络优化
network organization theory 网络组织理论
network organization structure 网络组织结构
network-oriented 面向网络的
network OS (network operating system) 网络操作系统
network outage 网络失效
network out dialing 网络向外拨号
network packet 网络封包,网络分组
network packet analysis 网络分组分析
network packet filtering framework 网络分组过滤防火墙
network packet monitoring 网络分组监视
network packet passed ratio 网络分组吞吐率
network packet-switching interface 网络分组交换接口
network parallel computing 网络并行计算
network parallel computing environment 网络并行计算环境

network paralysis 网络瘫痪
network parameter 网络参数
network parameter awareness 网络参数感知
network parameter control（NPC） 网络参数控制
network partition 网络分区
network partner plan 网络合作伙伴计划
network path 网络通路，网络路径
network payment 网络支付
network payment agency 网络支付中介
network payment platform 网络支付平台
network payment platform provider 网络支付平台提供商
network payment security 网络支付安全
network payment technologies 网络支付技术
network penetration 网络渗透
network penetration attack 网络渗透攻击
network penetration test 网路渗透测试
network performance analysis 网络性能分析
network performance analysis logical unit（NPALU） 网络性能分析逻辑单元
network performance parameter 网络性能参数
network peripheral 网络外围设备
network permission 网络权限
network phone 网络电话

network phone server 网络电话服务器
network physical separator 网络隔离器
network planning 网路规划
network planning implementation and promotion 网络策划执行与推广
network platform 网络平台
network point of access 网络接入点
network policy assembly 网络策略汇编
network port 网络端口
network port address translation 网络端口地址转换
network prefix 网络前缀
network prefix length 网络前缀长度
network prevention gateway（NPG） 网络防御网关
network printer 网络打印机
network print port 网络打印端口
network priority 网络优先级
network privacy 网络保密性
network problem determination aid 网络问题判定辅助程序
network problem determination application（NPDA） 网络问题判断应用程序
network processing engine 网络处理引擎
network processor 网络处理机
network product agent 网络产品代理
network product support（NPS） 网络产品支持
Network Professional Association（NPA） 网络专业人员协会
network promotion 网络推广

network promotion specialist 网络推广专员

network promotion strategy 网络推广战略

network propagation 网络传播

network propagation ethics 网络传播伦理

network protection schemes 网络保护机制

network protocol 网络协议

network protocol address information 网络协议地址信息

network protocol analysis 网络协议分析

network protocol analyzer 网络协议分析器

network protocol clear indication delay 网络协议拆线指示延迟

network protocol engine 网路协议引擎

network protocol simulation 网络协议仿真

network provider (NP) 网络提供商

network provider edge (NPE) 网络提供商边缘设备

network public opinion 网络舆情

network public opinion expression 网络民意表达

network public opinion guiding 网络舆论引导

network public opinion monitor 网络舆情监测

network public opinion monitoring system 网络舆情监控系统

network qualified name 网络验证名

network query language network query language

network raw data 网络原始数据

network reachability 网络可达性

network readiness testing 网络备用状态测试

network real time kinematic (NRTK) 网络实时动态

network recovery objective (NRO) 网络恢复目标

network redundancy 网络冗余

network reference service 网络咨询服务

network resource 网络资源

network resource directory 网络资源目录

network regulation 网络整治,网络调控

network resource management system 网络资源管理系统

network restoration ratio 网络恢复比

network restoration schemes 网络恢复机制

network robust 网络健壮性

network route 网络路由

network route control 网络路由控制

network rule 网络规则

network safety protection 网络安全防护

network sales promotion 网络促销

network scan attack 网络扫描攻击

network scan detection 网络扫描检测

network search engine 网络搜索引擎

network scenario emulation 网络场景仿真

network security 网络安全
network security analysis model 网络安全分析模型
network security awareness 网络安全意识
network security file system 网络安全文件系统
network security gateway 网络安全网关
network security holes 网络安全漏洞
network security isolation 网络安全隔离
network security isolation technique 网络安全隔离技术
network security scan system 网络安全扫描系统
network security separated card 网络安全隔离卡
network security protection 网络安全防护
network security protection system 网络安全防御系统
network security technology 网络安全技术
network security technology analysis 网络安全技术分析
network security technology platform 网络安全技术平台
network security transmission 网络安全传输
network security vulnerabilities 网络安全隐患
network segment 网段
network self-configuration 网络自主配置
network self-configuration and adaptive coordination 网络自主配置和自适应协调
network sensitivity 网络灵敏度
network separator 网络隔离器
network serial port adapter 网络串口适配器
network server 网络服务器
network service access point (NSAP) 网络服务接入点
network service engine 网络服务引擎
network service header 网络服务标题
network service location 网络服务定位
network service procedure error (NSPE) 网络服务过程错误
network service protocol (NSP) 网络服务协议
network service provider (NSP) 网络服务提供商
network service management 网络业务管理
network service procedure error (NSPE) 网络服务过程错误
network service sharing 网络服务分享
network service virtual connection (NSVC) 网络业务虚连接
network session 网络会话
network session accounting (NSA) 网络会话记账
network shared resource 网络共享的资源
network sharing 网络共享
network signaling 网络信令
network signaling load 网络信令负

载量
network signaling protocol (NSP) 网络信令协议
network signaling termination 网络信令终止
network signal transmission 网络信号传输
network simulation 网络仿真
network simulation parameter 网络仿真参数
network situational awareness 网络情境感知
network sink 网络汇流点,网络信宿
network size 网络规模
network slowdown 网络减速
network soccer game 网络足球游戏
network social 网络社会
network social behavio 网络社会行为
network social theory 网络社会理论
network software 网络软件
network space 网络空间
network stand-alone system 独立网络系统
network state awareness 网络状态感知
network storage server 网络存储服务器
network storage space 网络存储空间
network streaming 网络流
network streaming characteristic 网络流量特性
network streaming media 网络流媒体
network streaming media player 网络流媒体播放
network structure 网络结构

network structure parameter 网络结构参数
network supply load 网络负荷
network surveillance 网络监视
network survivability 网络生存性
network survivability function 网络生存性函数
network survivability performance 网络可存活性性能
network system structure 网络系统结构
network switch 网络切换
network switching cost 网络切换开销
network switching system 网络切换系统
network symbol 网络符号
network synchronization 网络同步
network synchronization unit 网络同步单元
network synchronous control 网络同步控制
network synchronous subsystem 网络同步子系统
network synchronous transmission 网络同步传输
network synthesis 网络综合
network system integration (NSI) 网络系统集成
network system planning 网络系统规划
network terminal protocol (NTP) 网络终端协议
network service server 网络服务服务器
network service system 网络服务系统

network terminating unit (NTU) 网络端接装置
network termination (NT) 网络终端
network termination equipment 网络终端设备
network termination type 1 (NT1) 一类网络终端
network termination type 2 (NT2) 二类网络终端
network that knows 网络能知
network theory 网络理论
network throughput 网络吞吐量
network time protocol (NTP) 网络时间协议
network time service 网络时间服务
network time synchronization 网络时间同步
network time synchronous protocol 网络时间同步协议
network timing 网络定时
network to network interface (NNI) 网络到网络接口
network token-ring interface 网络令牌环接口
network topology 网络拓扑
network topology construction 网络拓扑结构
network topology database 网络拓扑数据库
network topology structure 网络拓扑结构
network tort 网络侵权
network traffic 网络流量
network traffic content 网络流量内容
network traffic management system 网络业务管理系统
network traffic replay 网络流量回放
network trail termination point 网络跟踪终止点
network transfer function 网络传递函数
network transmission 网络传输
network transmission encryption 网络传输加密
network transmission medium 网络传输介质
network transparency 网络透明性
network trusted agent 网络可信代理
network trusted computing base 网络可信计算基
network trusted connection 网络可信连接
network-type virus 网络型病毒
network universality 网络通用性
network unsafe user 网络异常用户
network user address (NUA) 网络用户地址
network user identification (NUI) 网络用户标识
network user interface (NUI) 网络用户界面
network variable configuration table 网络变量配置表
network versus game 网络对战游戏
network video 网络视频
network video application 网络视频应用
network video broadcast 网络视频广播
network video multimedia streaming

网络视频多媒体流
network video on demand（NVOD） 网络视频点播
network video phone 网络视频电话
network video streaming 网络视频流
network video system 网络视频系统
network video transmission 网络视频传输
network violence 网络暴力
network virtual 网络虚拟
network virtual experiment 网络虚拟实验
network virtual file 网络虚拟文件
network virtual file system（NVFS） 网络虚拟文件系统
network virtual instrument 网络虚拟仪器
network virtualization 网络虚拟化
network virtual machine 网络虚拟机
network virtual property 网络虚拟财产
network virtual reality 网络虚拟现实
network virtual reality technology 网络虚拟现实技术
network virtual terminal（NVT） 网络虚拟终端
network virus 网络病毒
network virus monitoring 网络病毒监控
network virus scan 网络病毒扫描
network virus protection 网络病毒防范
network visible entity 网络可见实体
network visualization 网络可视化
network voice conference 网络语音会议
network voice protocol（NVP） 网络声音协议
network vulnerability 网络漏洞
network vulnerabilities scanner 网络漏洞扫描器
network vulnerability scanning 网络漏洞扫描
network weaving 网络编织
network windows system 网络窗口系统
network wiring services 网络布线服务，网络布线业务
network worm 网络蠕虫
network writer 网络写手
network zero knowledge proof system 网络零知识证明体制
network 3D virtual experiment 网络三维虚拟实验
NEU（network extension unit） 网络扩展单元
neural fuzzy fusion network 神经模糊融合网络
neural group network 神经元群网络
neural logic network 神经逻辑网络
neural network 神经元网络
neural network algorithm 神经元网络算法
neural network architecture 神经元网络体系结构
neural network diagnosis 神经元网络诊断
neural network fusion 神经元网络融合
neural network group 神经元网络组

neural network information fusion 神经元网络信息融合
neural network integration fusion model 神经元网络集成融合模型
neural network learning 神经元网络学习
neural network method 神经元网络法
neural network vertual reality 神经元网络虚拟现实
neural semantic network 神经元语义网络
newbie 菜鸟
news aggregation 新闻聚合
news aggregation services 新闻聚合服务
news aggregation site 新闻聚合网站
newsfeed 新闻传送干线
newsgroup 新闻组
NEXT（near-end cross talk） 近端串扰
next generation content delivery network 下一代内容发布网络
next generation firewalls 下一代防火墙
next generation Internet（NGI） 下一代因特网
next generation network（NGN） 下一代网络
next generation network focus group（NGNFG） 下一代网络焦点工作组
next generation operation support system（NGOSS） 下一代运营支持系统
next generation passive optical network（NG-PON） 下一代无源光网络
next generation service platform（NGSP） 下一代业务平台

next header 下一头部
next-hop forwarding 下一站转发
next hop resolution protocol（NHRP） 下一跳解析协议
next packet 下一个数据包
NFC（near field communication） 近场通信
NFM（network fault management） 网络故障管理
NFS（network file system） 网络文件系统
NGA（network gateway accounting） 网络网关记账
NGCP（network general control protocol） 网络通用控制协议
NGI（next generation Internet） 下一代因特网
NGIO（next generation input/output architecture） 下一代输入/输出结构
NGN（next generation network） 下一代网络
NGNFG（next generation network focus group） 下一代网络焦点工作组
NGOSS（next generation operation support system） 下一代运营支持系统
NG-PON（next generation passive optical network） 下一代无源光网络
NGSP（next generation service platform） 下一代业务平台
NHRP（next hop resolution protocol） 下一跳解析协议
NIC（network information center） 网络信息中心
NIC（network interface card） 网络接

口卡

NIC driver  网卡驱动程序

NICS (network information consulting service)  网络信息咨询服务

NIDL (network interface definition language)  网络接口定义语言

NIDS (network intrusion detection system)  网络入侵检测系统

NIE (network intelligence engine)  网络智能引擎

NII (national information infrastructure)  (美国)国家信息基础设施

NIM (network interface machine)  网络接口机

NIPS (network intrusion prevention system)  网络入侵预防系统

NIS (network information service)  网络信息服务

N-ISDN (narrow band integrated service digital network)  窄带综合业务数字网

NIST (National Institute of Standard and Technology)  (美国)国家标准和技术研究所

NLA (network logical address)  网络逻辑地址

NLDM (network logical data manager)  网络逻辑数据管理器

NLP (network layer protocol)  网络层协议

NLPI (network layer protocol identification)  网络层协议标识符

NLSP (NetWare link service protocol)  NetWare 链路服务协议

NLSP (network layer security protocol)  网络层安全协议

NMC (network management center)  网络管理中心

NMCS (network management consulting service)  网络管理咨询服务

NMF (Network Management Forum)  网络管理论坛

NMG (network management gateway)  网络管理网关

NMI (network node interface)  网络节点接口

NMIM (network management information model)  网络管理信息模型

NML (network management layer)  网络管理层

NMP (network management protocol)  网络管理协议

NMP (name management protocol)  名字管理协议

NMS (network management station)  网络管理站

NMS (network management system)  网络管理系统

NMSC (national mobile service center)  全国移动业务中心

NMVT (network management vector transport)  网络管理向量传输

NN (network node)  网络节点

NNCP (network node control point)  网络节点控制点

NNI (network to network interface)  网络到网络接口

NNI (network node interface)  网络节点接口

NNS (network node server)  网络节点

服务器
NNTP（network news transfer protocol） 网络新闻传递协议
NOC（network operation center） 网络运行中心
nodal attribute  节点属性
nodal constraint  节点限制
nodal function  节点功能
nodal metric  节点度量
nodal state parameter  节点状态参数
node  节点
node address  节点地址
node computer  节点计算机
node-edge weighted network  节点-边缘赋权网络
node encryption  节点加密
node identification  节点标识
node mobility  节点移动
node mobility model  节点移动模型
node name  节点名
node network  节点网络
node path control  节点路径控制
node probabilistic network  节点随机网络
node processor  节点处理器
node routing  节点路径选择
node type  节点类型
node verification  节点验证
noise network model  噪声网络模型
nominal bandwidth  额定带宽
nominal bit stuffing rate  标称比特填充率
non-adaptive algorithm  非自适应算法
non-associated CCIS  非结合式CCIS，非结合式公共信道局间信号传输

non-backtrace backward chaining  无回溯反向链接
non-bifurcated routing  单一路由
non-blocking interconnection network  无阻塞互连网络
non-blocking network  无阻塞网络
non-broadcast multiple access（NBMA）  非广播多路访问
non-classified signaling network  无级信令网
non-dedicated server  非专用服务器
non-deterministic network  随机网络
non-hierarchical requirement  无级选路网
non-native network  非本地网络
non-native network connection  非本地网络连接
non-persistent CSMA  非持续载波监听多路访问
non-persistent session  非持续会话
non-reala-time polling service（nrtPS）  非实时轮询服务
non-repudiation service  抗抵赖服务
non-SNA interconnection（NSI）  非系统网络体系结构互连
non-SNA station  非系统网络体系结构工作站
non-SNA terminal  非系统网络体系结构终端
non-source routed（NSR）  非源路由选择
non-standard ISDN terminal  非标准ISDN终端
non-switched point-to-point line  非交换点到点线路

non-transparent network communication 非透明网络通信
normal response mode (NRM) 正常响应方式
Nortel discovery protocol (NDP) 北电发现协议
North American Presentation Level Protocol Syntax (NAPLPS) 北美表示层协议语法
NOS (network operating system) 网络操作系统
NOSS (network operation support system) 网络运营支撑系统
notification channel (NCH) 通知信道
notification message 通知信息
notification window 通知窗口
Novell directory service (NDS) Novell 目录服务
Novell NetWare Novell 网络操作系统
Novell network 诺威网，Novell 网
Novell virtual terminal (NVT) Novell 虚拟终端
NOYB (none of your business) 与你无关
NP (network provider) 网络提供商
NPALU (network performance analysis logical unit) 网络性能分析逻辑单元
NPC (network parameter control) 网络参数控制
NPE (network provider edge) 网络提供商边缘设备
NPG (network prevention gateway) 网络防御网关
NPS (network product support) 网络产品支持
NPSI (X.25 NCP packet switching interface) X.25 网络控制程序分组交换接口
NREN (National Research and Education Network) （美国）国家研究和教育网
NRM (normal response mode) 正常响应方式
NRO (network recovery objective) 网络恢复目标
NRTK (network real time kinematic) 网络实时动态
NS (network service) 网络服务（程序）
NS (name server) 名字服务器
NS (neighbor solicitation) 邻居请求
NSA (network session accounting) 网络会话记账
NSAP (network service access point) 网络服务接入点
NSAPI (Netscape server application programming interface) 网景服务器应用编程接口
NSI (network system integration) 网络系统集成
NSI (non-SNA interconnection) 非系统网络体系结构互连
NSP (network service protocol) 网络服务协议
NSP (network signaling protocol) 网络信令协议
NSPE (network service procedure error) 网络服务过程错误
NSR (non-source routed) 非源路由选择

**NSS**（network and switching subsystem）网络交换子系统

**NSVC**（network service virtual connection）网络业务虚连接

**NT**（network termination） 网络终端

**n-terminal-pair network** n端对网络

**nTLDs**（national top level domain names） 国家顶级域名

**NTN**（national terminal number） 国家终端号

**NTONC**（National Transparent Optical Network Consortium） （美国）国家透明光网络联盟

**NTP**（network time protocol） 网络时间协议

**NTP**（network terminal protocol） 网络终端协议

**NTP amplification attack** 网络时间协议放大攻击，NTP放大攻击

**NTU**（network termination unit） 网络终端装置

**NT1**（network termination type 1） 一类网络终端

**NT2**（network termination type 2） 二类网络终端

**NUA**（network user address） 网络用户地址

**NUI**（network user identification） 网络用户标识

**NUI**（NetWare users international） NetWare 国际用户组织

**NUI**（network user interface） 网络用户界面

**null modem** 空调制解调器

**NVFS**（network virtual file system） 网络虚拟文件系统

**NVT**（network virtual terminal） 网络虚拟终端

**NVT**（Novell virtual terminal） Novell 虚拟终端

# O

OA（open access） 开放存取

OAA（open applications architecture） 开放应用体系结构

OAM（operation administration and maintenance） 运行、管理和维护

OAMC（operation administration and maintenance center） 运行、管理和维护中心

OAMC management function（OAMC-MF） OAMC管理功能

OAMC-MF（OAMC management function） OAMC管理功能

OAMC operation system（OAMC-OpS） OAMC运行系统

OAMC-OpS（OAMC operation system） OAMC运行系统

OAN（optical access network） 光接入网

OASIS（Organization for the Advancement of Structured Information Standards） 结构化信息标准推动组织

OBEX（object exchange） 对象交换

OBI（open buying on the internet） 互联网公开交易

object attribute 对象属性

object-based storage（OBS） 对象存储

object exchange（OBEX） 对象交换

object management 对象管理

object management group（OMG） 对象管理组

object middleware 对象中间件

object request brokers（ORB） 对象请求代理

OBSAI（open base station architecture initiative） 开放式基站架构

occupied bandwidth 占有带宽

OCDM（optical code division multiplexing） 光码分复用

OCDM（orthogonal code division multiplexing） 正交码分复用

OCDMA（optical code division multiple access） 光码分多址

OCS（optical circuit switching） 光路交换

ODA（open document architecture） 开放文档体系结构

ODAD（optimistic duplicate address detection） 乐观重复地址检测

ODBC（open database connectivity） 开放数据库互连

odd even check（OEC） 奇偶校验

odd even network 奇偶网络

odd even sorting 奇偶排序

odd even sorting network 奇偶排序网络

ODF（open document format） 开放文

档格式

ODI (open data-link interface) 开放数据链路接口

ODI/NDIS support (open data-link interface/network driver interface specification support) 开放数据链路接口/网络驱动接口规范的支持

ODINSUP (open data-link interface/network driver interface specification support) 开放数据链路接口/网络驱动接口规范的支持

ODMRP (on demand multicast routing protocol) 按需组播路由协议

ODN (optical distribution network) 光分配网

ODP (open directory project) 开放目录项目

ODS (open data server) 开放数据服务

ODSI (open directory service interface) 开放目录服务接口

ODSI (optical domain service interconnect) 光域业务互连

OEC (odd even check) 奇偶校验

OFBA (out of band authentication) 带外认证

OFC (open financial connectivity) 开放金融连接标准

OFDM (orthogonal frequency division multiplexing) 正交频分复用

OFDMA (orthogonal frequency division multiple access) 正交频分多址

office electronic mail 办公室电子邮件

office electronic mailbox 办公室电子邮箱

office local area network (OLAN) 办公室局域网

official node name 正式节点名字

official release code (ORC) 官方发布码

offline 离线,线下,网下

offline browser 离线浏览器

offline mail reader 离线电子邮件阅读器

offline navigator 离线浏览程序

offline player 离线播放器

offline to online to offline 线下营销到线上交易再到线下消费模式

OFNN (dynamic fuzzy neural network) 动态模糊神经网络

OFS (optical packet flow switching) 光分组流交换

OFX (open financial exchange) 开放式财务交换

OGSA (open grid service architecture) 开放网格服务体系结构

OGSI (open grid service infrastructure) 开放网格服务基础设施

OIC (only-in-chain) 链中唯一单元

OIC (Oh, I see) 哦,我知道

OIF (Optical Internetworking Forum) 光互联论坛

OIM (OSI Internet Management) 开放系统互连因特网管理组

OLAN (office local area network) 办公室局域网

OLT (optical line terminator) 光线路终端

OMA (Open Mobile Alliance) 开放移动联盟

OMC (operation and maintenance center)

操作维护中心

OME (open messaging environment) 开放消息传送环境

OMFI (open media framework interchange) 开放媒体架构互换

OMG (object management group) 对象管理组

OML (ontology markup language) 本体标记语言

OMP (open management protocol) 开放的管理协议

ONA (open network architecture) 开放性网络体系结构

ONC (open network computing) 开放网络计算

on demand paging 请求式页面调度

on demand services 点播业务

on demand multicast routing protocol (ODMRP) 按需组播路由协议

ONF (Open Network Foundation) 开放网络基金会

one-link webpage 单链接网页

one-point to multi-points communication 一点对多点通信

one-to-many non-blocking network 一对多无阻塞网络

one-to-many-link 一对多链接

one-to-one link 一对一链接

one way communication (OWC) 单向通信

one way message delay 单向报文延迟

one way only operation 单向工作

one way transmission 单向传输

onion routing (OR) 洋葱路由

online 网游,线上

online advertising 网络广告

online advertising alliance 网络广告联盟

online application service 在线预约申办服务

online auction purchasing 在线竞价采购

online bank 在线银行

online booking 在线预约,网络预约

online bookstore 网上书店

online casual game 在线游戏

online classified advertising 网络分类广告

online community 在线社区

online dating game 网上约会游戏

online direct mail advertising 线上直邮广告

online friendship 在线交友,网络交友

online game 在线游戏,网游

online games addiction 网络游戏成瘾

online games advertising 网络游戏广告

online hospital registration booking 网上医院预约挂号

online information service 在线信息服务

online library 网上图书馆

online marketing 在线营销,网络营销

online media advertising 网络媒体广告

online memory device 联机存储器件

online to offline (O to O,O2O) 线上到线下

online to offline commerce 线上到线下商业

online to place (O to P)　门店在线
online to partner (O2P)　在线合作伙伴
online payment　在线支付
online procurement　在线采购
online public opinion　网络舆论,网络舆情
online public opinion warning　网络舆情预警
online purchasing by invitation to bid　在线招标采购
online retailer　在线零售商
online service provider (OSP)　在线服务提供商
online service company　联机服务公司
online shopping　在线购物
online seller　在线卖家
online social media　网络社交媒体
online store　网上商店
online storefront　在线商店
online trading　网上交易
online video game　在线视频游戏
online virtual assets　网络虚拟财产
online virtual community　在线虚拟社区
online writer　在线作家
only a reward　打赏
only-in-chain (OIC)　链中唯一单元
ONN (open network node)　开放网络节点
on-off keying (OOK)　通断键控
ontology markup language (OML)　本体标记语言
OOF (out of frame)　帧失步
OOK (on-off keying)　通断键控
OoO (out of order)　乱序

OOT (object-oriented techniques)　面向对象技术
open access (OA)　开放存取
open applications architecture (OAA)　开放应用体系结构
open architecture　开放式系统结构,开放结构
open archival information system (OAIS)　开放档案信息系统
open base station architecture initiative (OBSAI)　开放式基站架构
open buying on the internet (OBI)　互联网公开交易
open collaboration network　开放式协作网络
open communication protocol　开放式通信协议
open computer network　开放式计算机网络
open core protocol　开放式核协议
open database access　开放式数据库存取
open database connectivity (ODBC)　开放数据库互连
open data center (ODC)　开放数据中心
Open Data Center Alliance (ODCA)　开放数据中心联盟
open data-link interface (ODI)　开放数据链路接口
open data-link interface/network driver interface specification support (ODINSUP, ODI/NDIS support)　开放数据链路接口/网络驱动接口规范的支持
open data server (ODS)　开放数据服务
open directory project (ODP)　开放目

录项目

open directory service interface (ODSI) 开放目录服务接口

open document architecture (ODA) 开放文档体系结构

open document format (ODF) 开放文档格式

open financial connectivity (OFC) 开放金融连接标准

open financial exchange (OFX) 开放式财务交换

OpenFlow protocol OpenFlow 协议

OpenFlow switch OpenFlow 交换机

OpenFlow Switch Consorlium OpenFlow 交换机论坛

OpenFlow technology OpenFlow 技术

Open Grid Forum (OGF) 开放网格论坛

open grid service architecture (OGSA) 开放网格服务体系结构

open grid service infrastructure (OGSI) 开放网格服务基础设施

open grid service system 开放网格服务系统

open license 开放式许可协议

open line protocol 开放式链接协议

open management protocol (OMP) 开放的管理协议

open media framework interchange (OMFI) 开放媒体架构互换

open messaging environment (OME) 开放消息传送环境

Open Mobile Alliance (OMA) 开放移动联盟

open MPEG consortium 开放 MPEG 协会

open network 开放网络

open network architecture (ONA) 开放性网络体系结构

open network computing (ONC) 开放网络计算

Open Network Foundation (ONF) 开放网络基金会

open network node (ONN) 开放网络节点

open profiling standard (OPS) 开放轮廓标准

open protocol 开放式协议

open protocol technology (OPT) 开放式协议技术

open service access (OSA) 开放业务接入

open service environment (OSE) 开放业务环境

open service gateway initiative (OSGI) 开放服务网关协议

open service interface definition (OSID) 开放服务接口定义

open shortest path first (OSPF) 开放最短路径优先

open shortest path first interior gateway protocol (OSPFIGP) 开放最短路径优先内部网关协议

open software descriptipon (OSD) 开放式软件描述

Open Software Foundation (OSF) 开放软件基金会

open source community 开源社区

open source data 开源信息,开源数据

open source development laboratory (OSDL)

开源发展实验室

Open Source Initiative (OSI) 开放源码促进会

open source license 开放许可证

open source software (OSS) 开放源代码软件

OpenStack 开放堆栈

open standard telecom architecture (OSTA) 开放式标准电信架构

open system architecture (OSA) 开放系统体系结构

open system environment (OSE) 开放系统环境

open information service environment 开放信息服务环境

open system interconnection (OSI) 开放系统互连

open system interconnection architecture (OSIA) 开放系统互连体系结构

open system interconnection security architecture (OSISA) 开放系统互连安全性体系结构

open system interconnection model 开放系统互连模型

open system interconnection/reference model (OSI/RM) 开放系统互连参考模型

open trading protocol (OTP) 开放交易协议

open transport network (OTN) 开放传输网络

open uniform resource locator (OpenURL) 开放统一资源定位符

OpenURL (open uniform resource locator) 开放统一资源定位符

operation administration and maintenance (OAM) 运行、管理和维护

operation administration and maintenance center (OAMC) 运行、管理和维护中心

operational support system (OSS) 运行支持系统

operation and maintenance center (OMC) 操作维护中心

operation system function (OSF) 运行系统功能

operations, administration and maintenance (OA&M) 操作、管理和维护

operator of broadband customer premises network 宽带用户驻地网运营商

operator border gateway router 运营商边界网关路由器

opportunistic network 机会网络

OPT (open protocol technology) 开放式协议技术

optical access network (OAN) 光接入网

optical add drop multiplexer (OADM) 光分插复用器

optical budget 光预算

optical burst switching (OBS) 光突发交换

optical burst switched network 光突发交换网络

optical bypass relay 光旁路中继器

optical carrier (OC) 光载波

optical carrier level 光载波级

optical circuit switching (OCS) 光路交换

optical clock distribution network 光时

钟分布网络

optical code division multiple access (OCDMA) 光码分多址

optical code division multiplexing (OCDM) 光码分复用

optical core network 光核心网

optical cross connector (OXC) 光交叉连接器

optical cross-connect panel 光交叉连接盘

optical distribution network (ODN) 光分配网

optical domain service interconnect (ODSI) 光域业务互连

optical fibre cable network 光纤有线网络

optical fiber channel 光纤信道

optical fiber communication 光纤通信

optical fiber communication network 光纤通信网

optical fiber sensor 光纤传感器

optical fiber sensor network 光纤传感器网络

optical fibre trunk network 光纤干线网络

optical fiber user network 光纤用户网

optical fibre user network system 光纤用户网系统

optical fiber user system 光纤用户系统

optical frequency division multiple access 光频多用复用接入

Optical Internetworking Forum (OIF) 光互联论坛

optical line terminator (OLT) 光线路终端

optical network 光网络, 光纤网络

optical network terminal (ONT) 光网络终端

optical network unit (ONU) 光网络单元

optical packet flow switching (OFS) 光分组流交换

optical packet switching (OPS) 光分组交换

optical packet switching network 光分组交换网络

optical path 光路, 光程

optical path automatic switching 光路自动切换

optical ring network 光环网

optical splitter 光分路器

optical supervisory channel (OSC) 光监控信道

optical switching (OS) 光交换

optical synchronous transport network 光同步传送网

optical time division multiplexing (OTDM) 光时分复用

optical transmission section (OTS) 光传输段

optical transmitter 光发射器

optical transport network (OTN) 光传送网

optical user network interface 光网用户网络接口

optical user terminal 光网用户终端

optical virtual private network (OVPN) 光虚拟专用网

optical waveguide technique 光波导

技术

optimal dropout compensator 最优丢失补偿器

optimistic duplicate address detection (ODAD) 乐观重复地址检测

optimum routing algorithm 最优路由算法

optimum routing selection 最优路径选择

opt-in E-mail marketing 单向确认邮件营销,选择式邮件营销

optional network facilities 选择网络设施

optional user facility 可选用户机制

option negotiation 选项协商

opt-out E-mail marketing 默认邮件营销

OR (onion routing) 洋葱路由

ORB (object request brokers) 对象请求代理

ORC (official release code) 官方发布码

ordering search 有序搜索

orderly release 顺序拆除

.org 组织机构域名

order path overhead monitor 低阶通道开销监视器

organic listing 自然排名

organizationally unique identifier (OUI) 机构唯一标识

Organization for the Advancement of Structured Information Standards (OASIS) 结构化信息标准推动组织

organization of network message 网络信息组织

organization network 组织网络,机构网络

origin address field prime 本地始端地址字段

originating UA 始发用户代理

origin element field (OEF) 起始单元字段

orthogonal code division multiplexing (OCDM) 正交码分复用

orthogonal frequency division multiple access (OFDMA) 正交频分多址

orthogonal frequency division multiplexing (OFDM) 正交频分复用

orthogonal transmission diversity (OTD) 正交发送分集

OSA (open systems architecture) 开放系统体系结构

OSD (open software descriptipon) 开放式软件描述

OSDL (open source development laboratory) 开源发展实验室

OSE (open service environment) 开放业务环境

OSE (open system environment) 开放系统环境

OSF (Open Software Foundation) 开放软件基金会

OSF (operation system function) 运行系统功能

OSGI (open service gateway initiative) 开放服务网关协议

OSI (open system interconnection) 开放系统互连

OSI (Open Source Initiative) 开放源码

促进会

OSIA (open system interconnection architecture) 开放系统互连体系结构

OSI base standards 开放系统互连基本标准

OSID (open service interface definition) 开放服务接口定义

OSI environment (OSIE) 开放系统互连环境

OSI functional standards 开放系统互连功能标准

OSI Internet management (OIM) 开放系统互连因特网管理组

OSI Level 1 开放系统互连的第一层
OSI Level 2 开放系统互连的第二层
OSI Level 3 开放系统互连的第三层
OSI Level 4 开放系统互连的第四层
OSI Level 5 开放系统互连的第五层
OSI Level 6 开放系统互连的第六层
OSI Level 7 开放系统互连的第七层

OSI network address 开放系统互连网络地址

OSI/Network Management Forum (OSI/NMF) 开放系统互连/网管论坛

OSI/NMF (OSI/Network Management Forum) 开放系统互连/网管论坛

OSI overall standards 开放系统互连总体标准

OSI presentation address 开放系统互连表示地址

OSI protocol 开放系统互连协议

OSI reference model standard 开放系统互连参考模型标准

OSI/RM (open system interconnection/reference model) 开放系统互连参考模型

OSISA (open system interconnection security architecture) 开放系统互连安全性体系结构

OSI service 开放系统互连服务

OSI service conventions 开放系统互连服务约定

OSI stack 开放系统互连协议栈

OSP (online service provider) 在线服务提供商

OSPF (open shortest path first) 开放最短路径优先

OSPFIGP (open shortest path first interior gateway protocol) 开放最短路径优先内部网关协议

OSS (open source software) 开放源代码软件

OSS (operational support system) 运行支撑系统

OSTA (open standard telecom architecture) 开放式标准电信架构

OTA (over the air) 空中下载

OTD (orthogonal transmission diversity) 正交发送分集

OTDM (optical time division multiplexing) 光时分复用

other-domain resource 非本域资源

OTN (open transport network) 开放传输网络

OTN (optical transport network) 光传送网

O to O (online to offline) 线上到线下

OTOH (on the other hand) 另一方面

O to P (online to place) 门店在线

OTP (open trading protocol) 开放交易协议
OTS (optical transmission section) 光传输段
OUI (organizationally unique identifier) 机构唯一标识
outbound link 导出链接
outbound path 输出通路
outbox 发件箱
outgoing access 出网访问
outgoing call 出网调用
out of band 带外
out of band authentication (OFBA) 带外认证
out-of-band signaling 带外信令
out of frame (OOF) 帧失步
output channel 输出通道
outside global address 外部全局地址
outside local address 外部本地地址
outside node 外部节点
over filled launch (OFL) 过满注入
over filled launch bandwidth (OFL-BW) 过满注入带宽
overflow route 溢出路由
overhead traffic 开销通信量
overlapping data channel 重叠数据通道
overlap signaling 重叠传信
overlay model 重叠模型
overlay network 覆盖网络
over sampling 过抽样
over the air (OTA) 空中下载
over the wall software 翻墙软件
OVPN (optical virtual private network) 光虚拟专用网
OWC (one way communication) 单向通信
OXC (optical cross connector) 光交叉连接器
O2O (online to offline) 线上对线下
O2P (online to partner) 在线合作伙伴

# P

P (peta) 帕,千万亿,千兆兆
PaaS (platform as a service) 平台即服务
PACCH (packet association control channel) 分组随路控制信道
pacing 调步
pacing group 调步组
pacing group size 调步组规模
pacing response 调步响应
pacing window 调步窗口
packed broadcast network 压缩广播网络
packed byte 压缩字节
packed data 压缩数据
packed data structure 压缩数据结构
packed field 压缩字段
packed format 压缩格式
packed format message 压缩格式的报文
packed key 压缩键标
packet 分组,包
packet access grant channel (PAGCH) 分组接入应答信道
packet addressing 分组寻址,包寻址
packet assembler/disassembler (PAD) 分装配/拆卸器
packet association control channel (PACCH) 分组随路控制信道
packet-based window scheme 基于分组的窗口模式
packet broadcast channel 分组广播信道
packet broadcast control channel (PBCCH) 分组广播控制信道
packet burst protocol (PBP) 分组突发协议
packet collision 分组碰撞,报文冲突
packet collision probability 报文冲突概率
packet collision reaction rate 碰撞反应速率
packet collision resolution 报文冲突分解
packet common control channel 分组公共控制信道
packet communication unit (PCU) 分组通信单元
packet control unit (PCU) 分组控制单元
packet data channel (PDCH) 分组数据信道
packet data control channel 分组数据控制信道
packet data protocol (PDP) 分组数据协议
packet data session 分组数据会话

packet data switching gateway　分组数据交换网关
packet data traffic channel（PDTCH）　分组数据业务信道
packet delay　分组延迟
packet details　数据包细节
packet dedicated control channel（PDCCH）　分组专用控制信道
packet discarding　分组丢弃
packet end control slot　分组尾控制时隙
packet error detection　分组差错检测
packet exchange protocol（PEP）　分组交换协议
packet filtering（PF）　分组过滤
packet filtering firewall　分组过滤防火墙
packet flow control　分组流控制
packet format　分组格式
packet header　分组头部
packet interleaving　分组交叉
packet Internet grope（PING）　乒，分组因特网探测器
packetized ensemble protocol（PEP）　分组总体协议
packet level　分组层
packet level filtering　分组层过滤
packet level interface　分组层接口
packet-level procedure（PLP）　分组层规程
packet lifetime control　分组生命周期控制
packet list　分组列表，数据包列表
packet loss　分组丢失，数据包丢失
packet loss compensating（PLC）　分组丢失补偿
packet loss concealment　分组丢失恢复
packet loss rate　分组丢失率
packet message delay　报文分组延迟
packet notify channel（PNCH）　分组通知信道
packet optical transport network（P-OTN）　分组光传送网
packet overhead　分组开销时间
packet over SONET　同步光纤网上分组传输技术
packet paging channel（PPCH）　分组寻呼信道
packet radio　分组交换无线电
packet radio network（PRnet）　分组无线网
packet radio network protocol　分组无线电网络协议
packet radio repeater（PRR）　分组无线转发器
packet radio station（PRS）　分组无线工作站
packet radio system（PRS）　分组无线电系统
packet radio terminal（PRT）　分组无线终端
packet radio unit（PRU）　分组无线设备
packet random access channel（PRACH）　分组随机接入信道
packet repeater　分组中继器
packet reservation multiple access（PRMA）　分组预留多址
packet reservation multiple access protocol　分组预约多址协议

packet retransmission interval 分组重发间隔
packet routing 分组路由选择
packet routing address 分组路由选择地址
packet satellite communication 分组卫星通信
packet sequence number 分组序号
packet sequencing 分组排序
packet service control agent function 分组服务控制代理功能
packet service control function 分组服务控制功能
packet service digital network 分组业务数字网
packet service gateway control function 分组业务网关控制功能
packet size 分组大小,包尺寸
packet slotted ring 分组时隙环
packet sniffer 包嗅探器
packet sniffing attack 包嗅探攻击
packets per second (PPS) 每秒分组数
packet switched bus 分组交换总线
packet switched channels 分组交换通道
packet switched connection type 分组交换连接类型
packet switched cross bar interconnect 分组交换纵横互联
packet switched data 分组交换数据
packet switched data network (PSDN) 分组交换数据网
packet switched mode 分组交换方式
packet switched network characteristic 分组交换网络特性
packet switched network tunnel 分组交换网络隧道
packet switched public data network 分组交换公用数据网
packet switched signaling message 分组交换信令报文
packet switched user traffic 分组交换用户通信量
packet switch equipment 分组交换设备
packet switching (PS) 分组交换
packet switching centre (PSC) 分组交换中心
packet switching data transmission service 分组交换数据传输服务
packet switching exchange (PSE) 分组交换机
packet switching network (PSN) 分组交换网络
packet switching node (PSN) 分组交换节点
packet switching service (PSS) 分组交换业务
packet switch interface unit 分组交换接口部件
packet switch level interface 分组交换级接口
packet switch module 分组交换模块
packet switch node 分组交换节点
packet switch stream (PSS) 分组交换流
packet timing control channel (PTCCH) 分组定时控制信道
packet transfer delay 分组传输时延
packet transfer protocol (PTP) 分组传送协议
packet type identifier 分组类型标识符
packet video protocol (PVP) 分组视频

协议
packet window 分组窗口
Pacific area communication network 太平洋地区通信网
Pacific area communication system 太平洋地区通信系统
Pacific area standards congress 太平洋地区标准会议
Pacific Asia Cooperative Telecommunications Network (PACTN) 亚太合作电信网
Pacific communications network 太平洋通信网络
Pacific Internet 太平洋因特网
Pacific ocean satellite 太平洋（通信）卫星
Pacific satellite 太平洋卫星
Pacific scatter system 太平洋散射系统
Pacific Telecommunication Council (PTC) 太平洋电信理事会
Pacific telecommunications conference 太平洋电信会议
PACTN (Pacific Asia Cooperative Telecommunications Network) 亚太合作电信网
PAD (packet assembler/disassembler) 分组装配/拆卸器
PAE (port access entity) 端口访问实体
PAGCH (packet access grant channel) 分组接入应答信道
page banner 页面旗帜广告
pagejacking 网页劫持
PageRank (PR) PR值，网页级别
page view (PV) 页面浏览量

page views per user 访问者的页面浏览数
paging channel (PCH) 寻呼信道
paging control channel (PCCH) 寻呼控制信道
paid listing 竞价排名
PAM (pulse amplitude modulation) 脉（冲）幅（度）调制
PAN (personal area network) 个人域网
PAN (parallel associative network) 并行联想网络
Pan African telecommunication network 泛非电信网
Pan European cellular digital mobile radio system 泛欧蜂窝数字移动无线电系统
Pan European cellular radio network 泛欧蜂窝无线电网
Pan European digital cellular land mobile telecommunication system 泛欧数字蜂窝移动通信系统
Pan European digital mobile communication 泛欧数字移动通信
PAP (port aggregation protocol) 端口聚集协议
PAP (password authentication protocol) 口令验证协议
PAP (push access protocol) 推送访问协议
PAR (positive acknowledgement with retransmission) 确定应答与重发
parallel and fault tolerant network 并行容错网络
parallel associative network (PAN) 并

行联想网络
parallel channel　并行通道
parallel communication　并行通信
parallel extended route　并行扩充路由
parallel fault tolerant routing algorithm　并行容错路由算法
parallel links　并行链路
parent peer group　上层平等组
partial availability trunk　部分可用性中继线
partial band jamming　部分频带干扰
partial break in echo suppressor　局部插入回波抑制器
partial broadcast channel　部分广播通道
partial carrier suppression　部分载波抑制
partial compaction　部分压缩
partial disturbed response signal　部分干扰响应信号
partial full duplex　部分全双工
partial packet discard　部分分组丢弃
partial redundancy　部分冗余
partial response code　部分响应代码
partial response coding　部分响应编码
partial response continuous phase modulation　部分响应连续相位调制
partial response line code　部分响应线路码
partial response modulation　部分响应调制
partial response signal　部分响应信号
partial response signaling system　部分响应信令系统
party line communication (PLC)　共线通信

passive attack　被动攻击
passive branched optical network　无源分支光网络
passive broadcast channel　无源广播信道
passive broadcast medium　无源广播媒体
passive bus　无源总线
passive circuit　无源电路
passive coaxial network　无源同轴电缆网络
passive communication monitoring　被动通信监控
passive communication satellite　无源通信卫星,被动式通信卫星
passive component　无源元[器]件
passive concentrator　无源集线器
passive detection　无源检测
passive double star　无源双星
passive duplicate address detection　无源重复地址检测,被动重复地址检测
passive eavesdropping　被动窃听
passive electronic card　无源电子卡
passive electronic countermeasure　无源电子对抗,消极电子对抗
passive electronic interception　被动电子截获
passive electronic receiver system　被动电子接收系统
passive fault detection　无源故障检测
passive FDM distributor　无源频分复用分配器
passive fiber component　光纤无源器件

passive gateway  无源网关
passive hub  无源集线器
passive microwave network  无源微波网络
passive network  无源网络
passive open  被动打开
passive optical coaxial hybrid network  无源光纤同轴混合网
passive optical device (POD)  无源光器件
passive optical distribution network (PODN)  无源光网络分配网
passive optical network (PON)  无源光网络
passive relay  被动中继,无源中继
passive relay station  无源中继台,无源中继站
passive star  无源星型连接
passive station  被动站,无源站
passive wiretapping  无源搭线窃听
pass-through  穿越
password authentication protocol (PAP)  口令验证协议
path analysis  路径分析
path asymmetric  路径不对称性
path attenuation  路径衰减
path attenuation device  通路衰减器
path control  路径控制
path control layer  路径控制层
path control network  路径控制网络
path cost  路径成本
path delay value  通路延迟值
path end point  通路端点
path expression  路径表达式
path field  通道域

path finding  路径寻找
path flag  路径标记
path frame  通路帧
path gain  通路增益
path generation method  路径产生方法,通路形成法
path independent insertion loss  通路无关插入损耗
path independent protocol  路径独立协议,与路径无关的协议
path information unit (PIU)  路径信息单元
path maximum transmission unit (PMTU)  路径最大传输单元
path maximum transmission unit (PMTU) discovery  路径最大传输单元发现
path monitoring (PM)  通道监视
path overhead (POH)  通道开销,路径负载
path sensitization  通路敏化
path trace  路径轨迹
path vector protocol  路径矢量协议
pay for performance (PFP)  按效果付费
payload length  负载长度
payload type  净荷类型
payload type indicator (PTI)  净荷类型指示符
payment gateway  支付网关
payment gateway certification center  支付网关认证中心
payment gateway interface  支付网关接口
payment with mobile phone  手机支付
pay per call (PPC)  按来电付费

pay per click (PPC)　点击付费
pay-per-view (PPV)　有偿收视服务
PBB (provider backbone bridge)　运营商骨干桥接
PBCCH (packet broadcast control channel)　分组广播控制信道
PBL (policy block list)　策略遏制列表
PBP (packet burst protocol)　分组突发协议
PBR (policy-based routing)　策略路由
PBT (provider backbone transport)　运营商骨干传送
PCCH (paging control channel)　寻呼控制信道
PCCPCH (primary common control physical channel)　主公共控制物理信道
PCF (point coordination function)　点协调功能
PCF interframe space (PIFS)　PCF帧间间隔
PCH (paging channel)　寻呼信道
PCI (protocol control information)　协议控制信息
PCM (pulse code modulation)　脉(冲编)码调制
PCN (personal communication network)　个人通信网络
PCPICH (primary common pilot channel)　基本公共导频信道
PCR (peak cell rate)　峰值信元速率
PCS (personal communication service)　个人通信业务
PCS (personal communication system)　个人通信系统
PCS (personal conference specification)　个人会议规范
PCS (physical coding sublayer)　物理编码子层
PCU (packet communication unit)　分组通信单元
PCU (packet control unit)　分组控制单元
PCU (peak concurrent users)　最高同时在线用户数
PCWG (personal conference working group)　个人会议工作组
PD (polarization diversity)　极化分集
PD (propagation delay)　传播时延
PDA (personal digital assistants)　个人数字助理
PDAU (physical delivery access unit)　物理投递访问单元
PDC (primary domain controller)　主域控制器
PDCCH (packet dedicated control channel)　分组专用控制信道
PDCH (packet data channel)　分组数据信道
.pdf (portable document format)　可移植文档文件名后缀
PDF (portable document format)　可移植文档格式
PDH (plesiochronous digital hierarchy)　准同步数字系列
PDN (public data network)　公用数据网
PDP (packet data protocol)　分组数据协议
PDP (packet data protocol)　分组数

协议

PDP (policy decision point) 策略决定点

PDS (premises distribution system) 综合布线系统

PDS (physical delivery system) 物理投递系统

PDSCH (physical downlink shared channel) 下行物理共享信道

PDTCH (packet data traffic channel) 分组数据业务信道

PDU (protocol data unit) 协议数据单元

PE (provider edge) 运营商边缘设备

peak cell rate (PCR) 峰值信元速率

peak concurrent users (PCU) 最高同时在线用户数

peak packet rate 峰值分组速率

peak transfer rate 最大传输率

PEAP (protected extensible authentication protocol) 受保护的可扩展认证协议

peer communication 对等通信

peer entities 对等实体

peer entity authentication 对等实体认证

peer group 对等组

peer group identifier 对等组标识符

peering 对等操作

peer layer 对等层

peer layer communication 对等层通信

peer model 对等模型

peer network 对等网络

peer node 对等节点

peer processes 对等进程

peer protocol 对等协议,同等协议

peer-to-peer (P2P) 对等

peer-to-peer architecture 对等体系结构

peer-to-peer chat network communication 网络上点对点聊天通信

peer-to-peer communication 对等体到对等体通信

peer-to-peer computing 对等计算

peer-to-peer file transfer 对等文件传送

peer-to-peer lending 网络借贷,点对点借贷

peer-to-peer network 对等网络

peer-to-peer overlay network 对等覆盖网络

peer-to-peer remote copy (PPRC) 对等远程拷贝

PEM (privacy enhanced mail) 保密增强邮件

PEP (packet exchange protocol) 分组交换协议

PEP (policy enforcement point) 策略增强点

PEP (packetized ensemble protocol) 分组总体协议

PER (proposed edited recommendation) 已修正的提议推荐

performance management (PM) 性能管理

peripheral border node 外设边界节点

peripheral link 外部链接

peripheral logical unit 外围逻辑单元

peripheral node 外部节点

peripheral path control 外部路径控制

peripheral physical unit  外部物理单元
permanent redirect  永久重定向
permanent virtual channel connection (PVCC)  永久虚拟通道连接
permanent virtual circuit (PVC)  永久虚拟线路
permanent virtual connection (PVC)  永久虚拟连接
permanent virtual path (PVP)  永久虚拟路径
permanent virtual path connection (PVPC)  永久虚拟路径连接
permissible code block  许可码组
permission E-mail marketing  许可式邮件营销
permit count  许可计数
permit flow control  许可流量控制
permit next increase (PNI)  允许下次增加
permit packet  许可数据包
per server licensing  每服务器许可协议
persistent CSMA  持续载波监听多路访问
persistent object  持久对象
persistent object identity  持久对象标识
persistent object layer  持久对象层
persistent object management  持久对象管理
persistent object service  持久对象服务
persistent object stores  持久对象存储
persistent route  持久路由
persistent session  持续会话
persistent URL (PURL)  持续统一资源定位符
personal Ad hoc network  个人自组织网络
personal area network (PAN)  个人域网
personal communication network (PCN)  个人通信网络
personal communication service (PCS)  个人通信业务
personal communication system (PCS)  个人通信系统
personal conference specification (PCS)  个人会议规范
personal conference working group (PCWG)  个人会议工作组
personal digital assistant (PDA)  个人数字助理
personal firewalls  个人防火墙
personal information bubble (PIB)  个人信息泡
personal mobile cloud computing (PMCC)  个人移动云计算
personal operating space (POS)  个人操作空间
personal Web server (PWS)  个人Web服务器
person-to-person service  个人至个人服务
pervasive computing  普适计算
petabyte (PB)  拍字节,千兆兆字节
pervasive computing  泛在计算,普适计算
pervasive computing environment  普适计算环境
PF (packet filtering)  分组过滤

PFP (pay for performance) 按效果付费
PGC (professional generated content) 专家创造内容
PGP (pretty good privacy) 优秀密钥
pharming 网址嫁接,域欺骗
pharming attack 域欺骗攻击
phase jitter 相位抖动
phase-locked loop quadrature phase shift keying (PLL-QPSK) 锁相环四相相移键控
phase modulation (PM) 相位调制
phase shift keying (PSK) 相移键控
phishing 网络钓鱼
phishing filter 钓鱼过滤器
phishing site 钓鱼网站
phone mail 电话邮件
photonic slot routing (PSR) 光子时隙路由选择
photonics switching 光交换
photonics switching technologies 光交换技术
physical channel 物理信道
physical coding sublayer (PCS) 物理编码子层
physical control layer 物理控制层
physical data layer 物理数据层
physical data link layer 物理数据链路层
physical data link network layer 物理数据链路网络层
physical delivery access unit (PDAU) 物理投递访问单元
physical delivery system (PDS) 物理投递系统

physical downlink shared channel (PDSCH) 下行物理共享信道
physical layer (PL) 物理层
physical layer connection 物理层连接
physical layer convergence protocol (PLCP) 物理层会聚协议
physical layer specification 物理层规范
physical layer protocol 物理层协议
physical level (X.25) 物理级(X.25)
physical link 物理链路
physical link layer 物理链路层
physical media dependent (PMD) 物理媒体相关
physical medium attachment (PMA) 物理媒体附件
physical medium attachment sublayer 物理媒体接触子层
physical medium dependent sublayer (PMDS) 物理媒体相关子层
physical medium layer 物理媒体层
physical metadata layer 物理元数据层
physical network 物理网络
physical network layer 物理网络层
physical random access channel (PRACH) 物理随机接入信道
physical service header (PSH) 物理服务头
physical signaling sublayer 物理信号子层
physical unit (PU) 物理单元
physical unit control point (PUCP) 物理单元控制点
physical unit service 物理单元服务
physical unit type 物理单元类型

physical uplink shared channel (PUSCH) 上行物理共享信道

PIB (personal information bubble) 个人信息泡

PIC (primary interexchange carrier) 主交换运营商

piconet 微微网

PICS (protocol implementation conformance statement) 协议实现一致性声明

PICS (platform for Internet content selection) 因特网内容选择平台

picture transfer protocol (PTP) 图片传输协议

PID (protocol identification) 协议标识符

PIDL (protocol identification descrimin language) 协议识别描述语言

PIFS (PCF interframe space) PCF 帧间间隔

piggyback acknowledgement 捎带应答

PIM (protocol independent multicast) 独立组播协议

PIM-DM (protocol independent multicast dense mode) 密集模式独立组播协议

PIM-SM (protocol independent multicast sparse mode) 稀疏模式独立组播协议

PING (packet Internet grope) 乒，分组因特网探测器

PING of death attack 乒死攻击

ping pong 乒乓

PIU (path information unit) 路径信息单位

PKC (public key cryptography) 公钥密码学

PKI (public key infrastructure) 公钥基础设施

PL (physical layer) 物理层

platform as a service (PaaS) 平台即服务

platform for Internet content selection (PICS) 因特网内容选择平台

platform for privacy preferences project (P3P) 隐私参数项目平台，P3P 协议

player versus environment (PVE) 环境对抗游戏

PLC (packet loss compensating) 丢包补偿

PLC (power line communication) 电力线通信

PLC (party line communication) 共线通信

plesiochronous digital hierarchy (PDH) 准同步数字系列

plex structure 丛结构

PLMN (public land mobile network) 公共陆地移动网

PLP (packet-level procedure) 分组级规程

plug-and-play networking 即插即用联网

PM (performance management) 性能管理

PM (phase modulation) 相位调制

PMA (physical medium attachment) 物理媒体附件

PMCC (personal mobile cloud computing) 人移动云计算

PMD (physical media dependent) 物理媒体相关

PMDS (physical medium dependent sublayer) 物理媒体相关子层

PMTU (path maximum transmission unit) 路径最大传输单元

PN (private network) 专用网

PNCH (packet notify channel) 分组通知信道

.png (portable network graphics) 可移植的网络图形文件名后缀

PNG datastream PNG 数据流,可移植的网络图形数据流

PNG decoder PNG 解码器,可移植的网络图形解码器

PNG editor PNG 编辑器,可移植的网络图形编辑器

PNI (permit next increase) 允许下次增加

PNNI (private network to network interface) 专用网间接口

PNNI protocol entity PNNI 协议实体,专用网间接口协议实体

PNNI routing control channel PNNI 路由控制通道,专用网间接口路由控制通道

PNP (private numbering plan) 专用网编号计划

POD (passive optical device) 无源光器件

PODA (priority oriented demand assignment) 面向优先级按需求分配

podcaster 播客

podcasting advertising 播客广告

podcasting marketing 播客营销

PODN (passive optical distribution network) 无源光网络分配网

POH (path overhead) 通道开销,路径负载

POI (point of Interface) 接口点

pointcast 点送,点播

pointcast network 点播网

point-cloud 点云

point-cloud data 点云数据

point-cloud compression 点云压缩

point-cloud fitting 点云拟合

point-cloud generation 点云生成

point coordination function (PCF) 点协调功能

point of interface (POI) 接口点

point of point (PoP) 接入点

point of presence (POP) 入网点

point of service (POS) 业务点

point-to-multipoint connection 点到多点连接

point to multi-point-group call (PTM-G) 点对多点群呼业务

point to multi-point multicast (PTM-M) 点对多点组播业务

point to multi-point service (PTM service) 点对多点业务

point-to-point channel-path configuration 点到点通道路径配置

point to point connectionless-mode network service (PTP-CLNS) 点对点无连接网络业务

point to point connection-mode network service (PTP-CONS) 点对点面向连接网络业务

point-to-point link 点对点链路

point-to-point network 点对点网络
point-to-point protocol (PPP) 点对点协议
point-to-point service 点对点业务
point-to-point topology 点对点拓扑
point-to-point transmission 点对点传输
point-to-point tunneling protocol (PPTP) 点对点隧道协议
poisoned reverse 毒性逆转
policy-based routing (PBR) 策略路由
policy block list (PBL) 策略遏制列表
policy decision point (PDP) 策略决定点
policy domain name resolution 策略域名解析
policy enforcement point (PEP) 策略增强点
polling 轮询
polling cycle 轮询周期
polling delay 轮询延迟
polling interval 轮询间隔
polling message 轮询信息
polling ratio 轮询比
PON (passive optical network) 无源光网络
P-OTN (packet optical transport network) 分组光传送网
PoP (point of point) 接入点
POP (point of presence) 入网点
POP (post office protocol) 邮局协议
pop-up ad 弹出式广告
pop-under ad 隐藏式弹出广告
portable network graphics (PNG) 可移植的网络图形

port access entity (PAE) 端口访问实体
port address translation (PAT) 端口地址转换
port aggregation protocol (PAP) 端口聚集协议
portal site 门户网站
port group 端口组
port identifier 端口标识符
port mirroring 端口镜像
port number 端口号
port sharing device 端口共享设备
port sharing unit (PSU) 端口共享部件
port switching hub 端口交换集线器
port trunking 端口聚合
port width 端口宽度
POS (point of service) 业务点
POS (personal operating space) 个人操作空间
POSIT (profiles for open systems internetworking technology) 开放系统互连网络技术概要
position determining entity (PDE) 定位实体
positive acknowledgement 肯定应答
positive acknowledgement with retransmission (PAR) 确定应答与重发
positive poll response 肯定查询响应
positive response 肯定应答
post 贴子
post bar 贴吧
post count 发帖量
posted 发贴
poster activity 发帖人活跃度
post office protocol (POP) 邮局协议

post office protocol version 3（POP3）邮局协议版本3

power line communication（PLC） 电力线通信

PPC（pay per click） 点击付费

PPC（pay per call） 按来电付费

PPCH（packet paging channel） 分组寻呼信道

PPDU（presentation protocol data unit）表示协议数据单元

p-persistent CSMA  p率持续载波监听多路访问

PPN（public packet network） 公用分组交换网

PPP（point-to-point protocol） 点对点协议

PPPoE（PPP over Ethernet） 以太网上的点对点协议

PPP over Ethernet（PPPoE） 以太网上的点对点协议

PPS（packets per second） 每秒分组数

PPS（prepaid service） 预付费业务

PPSN（public packet switched network）公用分组交换网

PPSS（public packet switching service）公用分组交换服务

PPTP（point-to-point tunneling protocol）点对点隧道协议

PQ（priority queue） 优先队列

PR（PageRank） PR值,网页级别

PR（proposed recommendation） 提议推荐

PR（pseudo range） 伪距

PRA（primary rate access） 基群速率接入

PRACH（packet random access channel）分组随机接入信道

PRACH（physical random access channel）物理随机接入信道

precise flow control 精细化流控

precision time protocol（PTP） 精确时间协议

precorrection 预校正

predictive coding 预测编码

preferential neighbor node 优先邻居节点

prefix-list 前缀列表

prefix renumbering 前缀重新编址

premises distribution system（PDS） 综合布线系统

prenegotiation phase 预协商阶段

prepaid service（PPS） 预付费业务

presentation layer 表示层

presentation protocol data unit（PPDU）表示协议数据单元

presentation service layer 表示服务层

pretty good privacy（PGP） 优秀密钥

primary authentication server 主认证服务器

primary common control physical channel（PCCPCH） 主公共控制物理信道

primary common pilot channel（PCPICH）基本公共导频信道

primary domain controller（PDC） 主域控制器

primary extended route 主扩充路由

primary group 基群

primary half-session 主会话端

primary interexchange carrier（PIC）主交换运营商

primary rate access (PRA)　基群速率接入
primary rate interface (PRI)　基群速率接口
primary route　主路由
primary session　基本会话
primary station　主站
priority oriented demand assignment (PODA)　面向优先级按需求分配
priority queue (PQ)　优先队列
privacy enhanced mail (PEM)　保密增强邮件
private ATM address　私有 ATM 地址
private circuit　专用线路
private cloud　私有云
private cloud computing　私有云计算
private cloud hosting　私有云托管
private cloud network　私有云网络
private cloud platform　私有云平台
private data network　专用数据网
private arbitrated loop　私有仲裁环路
private folders　专用文件夹
private identification number (PIN)　个人识别号
private network (PN)　专用网
private network to network interface (PNNI)　专用网间接口
private numbering plan (PNP)　专用网编号计划
private storage cloud　私有存储云
private switched network　专用交换网
private user network interface　专用用户网络接口
private virtual network (PVN)　专用虚拟网

private wire network　专线网络
PRMA (packet reservation multiple access)　分组预留多址
PRnet (packet radio network)　分组无线网
proactive routing　先验式路由选择
probabilistic Boolean network　概率布尔型网络
probabilistic inference network　概率推断网络
probabilistic neural network　概率神经网络
probabilistic reasoning network　概率推理网络
ProBlogger　职业博客,专业博客
process neural network　过程神经元网络
professional generated content (UGC)　专家创造内容
professional service website search engine　专业服务网站搜索引擎
profiles for open systems internetworking technology (POSIT)　开放系统互连网络技术概要
promiscuous mode　混杂模式
promiscuous-mode transfer　混杂式传送
propagation delay (PD)　传播时延
proportional fair scheduling (PFS)　正比公平调度
proposed edited recommendation (PER)　已修正的提议推荐
proposed parameter　建议参数
proposed recommendation (PR)　提议推荐

proprietor of network  网络所有者
protected extensible authentication protocol (PEAP)  受保护的可扩展认证协议
protocol  协议
protocol adapter  协议适配器
protocol address  协议地址
protocol analyzer  协议分析器
protocol boundary  协议边界
protocol configuration  协议配置
protocol control  协议控制
protocol control information (PCI)  协议控制信息
protocol conversion  协议转换
protocol converter  协议转换器
protocol data unit (PDU)  协议数据单元
protocol emulator  协议仿真器
protocol engineering  协议工程
protocol entity  协议实体
protocol identification (PID)  协议标识符
protocol identification module  协议标识符模块
protocol identification descrimion language (PIDL)  协议识别描述语言
protocol implementation conformance statement (PICS)  协议实现一致性声明
protocol independent multicast (PIM)  独立组播协议
protocol independent multicast dense mode (PIM-DM)  密集模式独立组播协议
protocol independent multicast sparse mode (PIM-SM)  稀疏模式独立组播协议
protocol insensitive  协议不敏感
protocol isolation  协议隔离
protocol levels  协议级
protocol machine  协议机
protocol port  协议端口
protocol port number  协议端口号
protocol stack  协议堆栈
protocol translator  协议翻译器
provider backbone bridge (PBB)  运营商骨干桥接
provider backbone bridged network  运营商骨干桥接网络
provider backbone transport (PBT)  运营商骨干传送
provider bridged network  运营商网桥网络
provider edge (PE)  运营商边缘设备
provider entity  供应实体
proxy ARP  代理 ARP,代理地址解析协议
proxy address resolution protocol  代理地址解析协议
proxy cache  代理缓存
proxy cache server  代理缓存服务器
proxy gateway  代理网关
proxy server  代理服务器
proxy server address  代理服务器地址
proxy server list  代理服务器列表
proxy server model  代理服务器模型
proxy server setup  代理服务器设置
PRR (packet radio repeater)  分组无线转发器

PRS (packet radio system) 分组无线电系统
PRT (packet radio terminal) 分组无线终端
PRU (packet radio unit) 分组无线设备
PSC (packet switching centre) 分组交换中心
PSDN (packet switched data network) 分组交换数据网
PSE (packet switching exchange) 分组交换机
pseudo-address 伪地址
pseudo-flaw 伪缺陷
pseudo object 伪对象
pseudo range (PR) 伪距
pseudo wire (PW) 伪线
pseudo wire emulation edge-to-edge (PWE3) 边缘到边缘的伪线仿真
PSH (physical service header) 物理服务头
PSK (phase shift keying) 相移键控
PSN (public switching network) 公用交换网络
PSN (packet switching network) 分组交换网络
PSN (packet switching node) 分组交换节点
PSR (photonic slot routing) 光子时隙路由
PSS (packet switching service) 分组交换业务
PSS (packet switch stream) 分组交换流
PSTN (public switched telephone network) 公用交换电话网
PTCCH (packet timing control channel) 分组定时控制信道
PTM-G (point to multi-point-group call) 点对多点群呼业务
PTM-M (point to multi-point multicast) 点对多点组播业务
PTN (public telephone network) 公共电话网
PTP (packet transfer protocol) 分组传送协议
PTP (picture transfer protocol) 图片传输协议
PTP (precision time protocol) 精确时间协议
PTP-CLNS (point to point connectionless-mode network service) 点对点无连接网络业务
PTSE (PNNI topology state element) PNNI拓扑状态元素,专用网间接口拓扑状态元素
PTSP (PNNI topology state packet) PNNI拓扑状态包,专用网间接口拓扑状态包
PU (physical unit) 物理单元
Public Blockchains 公有区块链
public cloud 公有云,公共云
public cloud platform 公共云平台
public cloud service 公共云服务
public cloud storage 公共云存储
public data network (PDN) 公用数据网
public data transmission service 公用数据传输服务
public file 公共文件

public folders 公用文件夹
public key cryptography (PKC) 公钥密码学
public key crypto system 公钥密码系统
public key infrastructure (PKI) 公钥基础设施
public land mobile network (PLMN) 公共陆地移动网
public libraries website 公共图书馆网站
public mailing list 公共邮件发送清单
public message 公用消息
public network 公用网络
public opinion in Web 网络舆论
public packet network (PPN) 公用分组网络
public packet switched network (PPSN) 公用分组交换网
public packet switching service (PPSS) 公用分组交换服务
public policy website 公共政策网站
public standard protocol 公共标准协议
public security network 公安网
public switched network (PSN) 公用交换网络
public switched telephone network (PSTN) 公用交换电话网
public telephone network (PTN) 公共电话网
Public Web 公共网络
public wireless local area network (PWLAN) 公共无线局域网
PUCP (physical unit control point) 物理单元控制点
pull media 拖拉媒体
pull technology 拖拉技术
pulse amplitude modulation (PAM) 脉(冲)幅(度)调制
pulse code modulation (PCM) 脉(冲编)码调制
pulse input asynchronous network 脉冲输入异步网络
pulsing zombie 脉动僵尸机
pure carrier network 纯载波网络
PURL (persistent URL) 持续统一资源定位符
PUSCH (physical uplink shared channel) 上行物理共享信道
push access protocol (PAP) 推送访问协议
push client 推送客户器
push media 推送媒体
push technology 推送技术
PVC (permanent virtual circuit) 永久虚拟线路
PVC (permanent virtual connection) 永久虚拟连接
PVCC (permanent virtual channel connection) 永久虚拟通道连接
PVE (player versus environment) 环境对抗游戏
PVN (private virtual network) 专用虚拟网
PVP (packet video protocol) 分组视频协议
PVP (permanent virtual path) 永久虚拟路径
PVPC (permanent virtual path connection)

永久虚拟路径连接
**PW**（pseudo wire） 伪线
**PWE3**（pseudo wire emulation edge-to-edge） 边缘到边缘的伪线仿真
**PWLAN**（public wireless local area network） 公共无线局域网
**PWS**（personal Web server） 个人 Web 服务器
**pyramid configuration** 金字塔型配置
**P2P**（peer-to-peer） 对等
**P2P lending** 点对点借贷，网络借贷
**P2P network** 对等网络
**P3P**（platform for privacy preferences project） 隐私参数项目平台

# Q

Q (quality factor)  品质因数
QA (quality assurance)  质量保证
QA (Q-adapter)  Q 适配器
Q-adapter (QA)  Q 适配器
Q-adapter function (QAF) block  Q 适配器功能块
QAF (Q-adapter function)  Q 适配器功能
QAM (quadrature amplitude modulation)  正交幅度调制
QC (quality control)  质量控制
QCI (QoS class identifier)  服务质量类别标识
Q interface  Q 接口
QMTP (quick mail transfer protocol)  快速邮件传送协议
QoE (quality of experience)  体验质量
QoS (Quality of Service)  服务质量
QoS class identifier (QCI)  服务质量类别标识
Q signal  Q 信号
quadrature amplitude modulation (QAM)  正交幅度调制
qualified logical link control (QLLC)  限定式逻辑链路控制
qualified logical link control packet assembler disassembler (QLLC PAD)  限定式逻辑链路控制分组装拆设备

quality assurance (QA)  质量保证
quality control (QC)  质量控制
quality of experience (QoE)  体验质量
quality of service (QoS)  服务质量
quality of service maintenance  服务质量维护
quality of service negotiation  服务质量协商
quality of service renegotiation  服务质量重协商
quantitative self-organization neural network  量化自组织神经网络
quantum cellular neural network  量子细胞神经网络
quantum communication network  量子通信网络
quantum computing network  量子计算网络
quantum direct communication  量子直接通信
quantum entangled  量子纠缠
quantum entangled distribution  量子纠缠分发
quantum entangled communication  量子纠缠通信
quantum entangled information  量子纠缠信息
quantum entangled state  量子纠缠态

quantum genetic and quantum neural network hybrid algorithm 量子遗传神经网络混合算法

quantum information network 量子信息网络

quantum interference network 量子相干网络

quantum key distribution system 量子密钥分发系统

quantum network 量子网络

quantum network communication 量子网络通信

quantum neural computational network 量子神经计算网络

quantum private communication 量子保密通信

quantum repeater 量子中继站

quantum satellite 量子卫星

quantum secure communication 量子安全通信

quantum synchronous communication 量子同步通信

quarantine service 隔离服务

quasi-associated mode 准随路方式

queue-based share 基于队列的共享

queued printing service 排队打印服务

queuing delay 排队延迟

queuing delay time 排队延迟时间

queuing network model 排队网络模型

queuing problem 排队问题

quick mail transfer protocol (QMTP) 快速邮件传送协议

quoting 引用

Q3 interface Q3 接口

Q.921 Q.921 标准, ISDN 用户网络接口数据链路第二层协议

Q.931 Q.931 标准, ISDN 用户网络接口数据链路第三层协议

# R

RA(routing area) 路由区
RA(router advertisement) 路由器通告
RAC(radio admission control) 无线接纳控制
RACH(random access channel) 随机接入信道
RACF(resource admission control function) 资源接纳控制功能
RACH(random access channel) 随机接入信道
RACL(reflexive access control list) 自反访问控制列表
RACS(resource admission control subsystem) 资源接纳控制子系统
radio access network(RAN) 无线接入网
radio access network application part(RANAP) 无线接入网络应用部分
radio admission control(RAC) 无线接纳控制
radio base station(RBS) 无线基站
radio bearer control(RBC) 无线承载控制
radio communication network 无线通信网
radio frequency for consumer electronics(RF4CE) 消费电子射频

radio frequency identification(RFID) 射频识别
radio link(RL) 无线链路
radio link set(RLS) 无线链路集
radio link control(RLC) 无线链路控制
radio link protocol(RLP) 无线电链路协议
radio multiplexing 无线电多路复用
radio network 无线网络
radio network controller(RNC) 无线网络控制器
radio network planning 无线网络规划
radio network subsystem(RNS) 无线网络子系统
radio network temporary identity(RNTI) 无线网络临时身份
radio resource control(RRC) 无线资源控制
radio resource management(RRM) 无线资源管理
radio resource service access point(RRSAP) 无线资源业务接入点
RADIUS(remote authentication dial-in user service) 远程认证拨号用户服务
RADSL(rate adaptive digital subscriber link) 速率自适应数字用户线路

RAIN (regional access information network) 地区接入信息网
RAN (radio access network) 无线接入网
RANAP (radio access network application part) 无线接入网络应用部分
random access channel (RACH) 随机接入信道
random access method 随机访问方式
random access termination 随机接入终端
random access termination identifier (RATI) 随机接入终端标识
random assignment multiple access (RSMA) 随机分配多址
random multiple access 随机多重访问
random multiple access communication (RMAC) 随机多路存取通信,随机多址通信
random multiple access protocol 随机多址接入协议
random sequential access 随机顺序存取
random walk routing (RWR) 随机游走路由选择
rapid cloud transmission (RCT) 极速云传输
RAR (route addition resistance) 路由附加阻力
RARP (reverse address resolution protocol) 反向地址解析协议
RARP server 反向地址解析协议服务器
RAS (remote access service) 远程访问服务
RAS (remote access server) 远程访问服务器
rate adaptation 速率适配
rate adaptive digital subscriber link (RADSL) 速率自适应数字用户线路
rate decrease factor (RDF) 速率降低因子
RATI (random access termination identifier) 随机接入终端标识
RAU (remote radio unit) 远端射频单元
raw socket 原始套接字
RBGAN (regional broadband global area network) 区域性宽带全球网
RBS (radio base station) 无线基站
RCC (route control center) 路由控制中心
RCT (rapid cloud transmission) 极速云传输
RD (router discovery) 路由器发现
RD (routing domain) 路由选择域
RDF (resource description framework) 资源描述框架
RDF (record definition field) 记录定义字段
RDF (rate decrease factor) 速率降低因子
RDFS (resource description framework schema) 资源描述框架模式
RDG (routing domain group) 路由选择域组
RDI (routing domain identifier) 路由选择域标识符
RDP (reliable datagram protocol) 可靠

数据报协议
reactive routing 反应式路由
really simple syndication (RSS) 简易信息聚合，聚合内容
real network address 网络实地址
real open system 真正开放系统
real time available bit rate 实时可用位速率
real time control 实时控制
real time messaging protocol (RTMP) 实时消息传送协议
real time polling service (rtPS) 实时轮询服务
real time search 实时搜索
real time service flow measurement (RTFM) 实时业务流测量
real time streaming protocol (RTSP) 实时流媒体协议
real time transport control protocol (RTCP) 实时传送控制协议
real time transport protocol (RTP) 实时传送协议
rearrangeable non-blocking network 可重排无阻塞网络
reasoning network 推理网络
reassembly 重装，重组
reassembly deadlock 重组死锁
reassembly-free deep packet inspection (RFDPI) 免重组深度数据包检测
reassembly timer 重组定时器
reboot 重新启动
receive diversity 接收分集
receiving window 接收窗口
reception congestion 接收拥塞
reciprocal link 交互链接

reconciliation sublayer (RS) 协调子层
reconfigurable optical add drop multiplexer (ROADM) 可重构光分插复用器
record protocol layer 记录协议层
record route option 路由记录选项
recurrent fuzzy neural network 递归模糊神经网络
recurrent multilayer neural network 多层递归神经网络
recurrent process neural network 递归过程神经网络
recurrent wavelet neural network (RWNN) 递归小波神经网络
recurrent neural network 递归神经网络
recurrent neural network with bias unit 带有偏差单元的递归神经网络
red envelope 红包
redirector 重新定向
redistribution 重新分配
reference configuration 参考配置
reference point 基准点
reflect attack 反射攻击
reflexive access control list (RACL) 自反访问控制列表
regenerator section overhead (RSOH) 再生段开销
regional access information network (RAIN) 地区接入信息网
regional broadband global area network (RBGAN) 区域性宽带全球网
regional computer network 区域计算机网
regional foreign agent (RFA) 区域外

地代理

Regional Internet Register (RIR)　地区性因特网注册机构

regional signaling transfer points (RSTP)　大局信令中转点

registered ports　注册端口

register insertion　寄存器插入法

register insertion ring　寄存器插入环

register of known spam operations (ROKSO)　已知垃圾邮件运营者

register signaling　记发器信令

registrable resource　可登记资源

registration　注册

registration relationship　注册关系

regular fuzzy neural network　正则模糊神经网络

relative URL　相对 URL,统一资源标识符

relay-enhanced cellular network　中继增强蜂窝网络

relay message　中继消息

relaynet international message exchange (RIME)　中继网国际消息交流

relay open system　中继开放系统

reliable datagram protocol (RDP)　可靠数据报协议

reliable stream protocol (RSP)　可靠流协议

reliable transfer service element (RTSE)　可靠传输服务元素

remote access data processing network　远程访问数据处理网络

remote access server (RAS)　远程访问服务器

remote access service (RAS)　远程访问服务

remote assistance　远程维护

remote authentication dial-in user service (RADIUS)　远程认证拨号用户服务

remote computer network　远程计算机网

remote control network　远程控制网络

remote earth station　远地地面站

remote file service (RFS)　远程文件业务

remote job entry protocol (RJEP)　远程作业录入协议

remote login　远程登录

remote loop adapter　远程环路适配器

remote mail gateway host　远端邮件网关主机

remote method invocation (RMI)　远程方法调用

remote mirroring　远程镜像

remote name server　远程名字服务器

remote network　远程网络

remote network access (RNA)　远程网络访问

remote network closed loop control　远程网络闭环控制

remote operation service element (ROSE)　远程操作服务元素

remote procedure call (RPC)　远程过程调用

remote radio unit (RAU)　远端射频单元

remote resource　远程资源

remote resource access validation　远程资源访问验证

remote station　远程站

remote sub station 远程子站
remote technical assistance and information network 远程技术援助与信息网络
remote work station 远程工作站
rendering engine 渲染引擎
rendezvous point (RP) 汇聚点
REP (robot exclusion protocol) 机器人排除协议
repeater (RP) 中继器
repeat visitor (RV) 重复访客
replay attack 重放攻击
reply post 回贴,跟贴
representation medium 表示媒体
repudiation 否认
request disconnect (RD) 请求断开
request for comments (RFC) 请求注解
request/response header (RH) 请求/应答标题
request/response unit (RU) 请求/应答单元
request to send/clear to send (RTS/CTS) 请求发送/清除发送协议
request unit (RU) 请求单元
Reseaux IP Europeens (RIPE) 欧洲 IP 地址注册中心
reset collision 复位冲突
reset-confirmation packet 复位确认包
resilient fast Ethernet ring (RFER) 弹性快速以太网环路
resilient packet ring (RPR) 弹性分组环
resource admission control function (RACF) 资源接纳控制功能
resource admission control subsystem (RACS) 资源接纳控制子系统
resource description framework (RDF) 资源描述框架
resource description framework schema (RDFS) 资源描述框架模式
resource island 资源孤岛
resource hierarchy 资源体系
resource interchange file format (RIFF) 资源交换文档格式
resource management (RM) 资源管理
resource management cell 资源管理信元
resource registration 资源登记
resource reservation protocol (RSVP) 资源预留协议
resource routing 资源路由选择
resource sharing network 资源共享网络
resource sharing network platform 资源共享网络平台
response time 响应时间
response time window 响应时间窗口
response unit (RU) 应答单元
restricted subnetwork 受限子网
return link 返回链接
reverse address resolution protocol (RARP) 反向地址解析协议
reverse caching 反向缓存
reverse-path broadcasting 反向路径广播
reverse path forward (RPF) 逆向路径转发
reverse proxy 反向代理
RFA (regional foreign agent) 区域外地代理

RF-based mobile payment 基于射频的移动支付
RFC (request for comments) 请求注解
RFDPI (reassembly-free deep packet inspection) 免重组深度数据包检测
RH (request/response header) 请求/应答标题
RIB (routing information base) 路由选择信息库
rich E-mail 增强式电子邮件
rich media 富媒体
rich media cloud 彩云
RID (resource identifier) 资源标识符
RIF (routing information field) 路由信息段
RIFF (resource interchange file format) 资源交换文档格式
RIME (relaynet international message exchange) 中继网国际消息交流
ring attaching device 环形网附接设备
ring communication network 环形通信网络
ring connection 环形连接
ring control network 环形控制网
ring diagnostic 环路诊断
ring group 环组
ring interface adapter 环形接口适配器
ring interworking 环互通
ring local network 环形局部网络
ring network 环形网络
ring neural network 环形神经网络
ring network node 环网节点
ring network structure 环网结构
ring station 环站

ring switched computer network 环形交换计算机网络
ring topology 环形拓扑
ring wiring concentrator 环接线集中器
R interface R 接口
RIP (routing information protocol) 路由信息协议
RIP addressing conventions RIP 编址约定，路由信息协议编址约定
RIP algorithm rules RIP 算法规则，路由信息协议算法规则
RIPE (Reseaux IP Europeens) 欧洲 IP 地址注册中心
RIR (Regional Internet Register) 地区性因特网注册机构
RKRL (radio knowledge representation language) 无线电知识描述语言
RJ-45 connector RJ-45 连接器
RL (radio link) 无线链路
RLC (radio link control) 无线链路控制
RLP (radio link protocol) 无线电链路协议
RLS (radio link set) 无线链路集
RM (resource management) 资源管理
RMAC (random multiple access communication) 随机多路存取通信，随机多址通信
RMI (remote method invocation) 远程方法调用
RN (resource number) 资源号
RNA (remote network access) 远程网络访问
RNAA (request network address assignment)

请求网络地址赋值
RNC (radio network controller)　无线网络控制器
RNS (radio network subsystem)　无线网络子系统
RNTI (radio network temporary identity)　无线网络临时身份
ROADM (reconfigurable optical add drop multiplexer)　可重构光分插复用器
roaming　漫游
robot exclusion protocol (REP)　机器人排除协议
robust header compression (ROHC)　强健包头压缩
robust knowledge-based neural network model　稳健的知识神经网络模型
robust network　稳健网络
robust neural network　稳健神经网络，鲁棒神经网络
robust security network (RSN)　强健安全网络
robust security network association (RSNA)　强健安全网络组合
ROHC (robust header compression)　强健包头压缩
ROKSO (register of known spam operations)　已知垃圾邮件运营者
role-based access control (RBAC)　基于角色的访问控制
rollover cable　反转线
role interaction network　角色互动网络
root application context　根应用上下文
root bridge　根网桥
root port　根端口
root server　根服务器

root server system　根服务器系统
Root Server System Advisory Committee (RSSAC)　根服务器系统咨询委员会
round-trip propagation time　往返传播时间
round-trip time (RTT)　往返时间
route　路由
route addition resistance (RAR)　路由附加阻力
route aggregation　路由聚合，路由汇聚
route attack　路由攻击
route control center (RCC)　路由控制中心
route convergence　路由收敛性
route discovery　路由探索
route extension (REX)　路由扩展件
route flap　路由抖动
route learning　路由学习
route matrix　路由矩阵
route optimization　路由最佳化
route processor　路由处理器
router　路由器
router advertisement (RA)　路由器通告
router discovery (RD)　路由器发现
router on a stick　单臂路由器
router solicitation (RS)　路由器请求
route selection　路由选择
route selection control vector (RSCV)　路由选择控制向量
route selection service (RSS)　路由选择服务
route server　路由服务器
route switching　路由交换，路由转发
route switching policy　路由转发策略
route switching subsystem　路由交换子系统

route table generator (RTG) 路由表生成器
route update packet 路由更新包
route weight 路由权重
routing 路由选择
routing algorithms 路由选择算法
routing alternate 可替换路由选择
routing and wavelength assignment (RWA) 路由和波长分配
routing and wavelength assignment algorithm 路由和波长分配算法
routing area (RA) 路由区
routing bridge 路由桥
routing by destination 按目的地选择路由
routing by key 按键标选择路由
routing code 路由选择码
routing control 路由选择控制
routing convergence 路由收敛
routing convergence time 路由收敛时间
routing domain (RD) 路由选择域
routing domain group (RDG) 路由选择域组
routing domain identifier (RDI) 路由选择域标识符
routing indicator 路由选择指示符
routing information base (RIB) 路由选择信息库
routing information field (RIF) 路由信息段
routing information protocol (RIP) 路由信息协议
routing loop 路由环路
routing matrix 路由矩阵
routing metric 路由选择量度
routing problem 路由问题
routing protocol 路由协议
routing qualifier 路由检验器
routing queue 路由队列
routing restriction 路由限制
routing strategies of packet radio network 分组无线网络路由选择策略
routing table 路由表
routing table maintenance protocol (RTMP) 路由表维护协议
routing update 路由选择更新
RPF (reverse path forward) 逆向路径转发
RPR (resilient packet ring) 弹性分组环
RRC (radio resource control) 无线资源控制
RRM (radio resource management) 无线资源管理
RRSAP (radio resource service access point) 无线资源业务接入点
RS (reconciliation sublayer) 协调子层
RSCV (route selection control vector) 路由选择控制向量
RSN (robust security network) 强健安全网络
RSNA (robust security network association) 强健安全网络组合
RSOH (regenerator section overhead) 再生段开销
RSP (reliable stream protocol) 可靠流协议
RSS (route selection service) 路由选择服务
RSS (really simple syndication) 简易信息聚合,聚合内容
RSSAC (Root Server System Advisory

Committee) 根服务器系统咨询委员会
RSS reader 聚合内容阅读器
RST (regenerator section termination) 再生段终端
RSTP (regional signaling transfer points) 大局信令中转点
RSVP (resource reservation protocol) 资源预留协议
RTCP (real time transport control protocol) 实时传送控制协议
RTFM (real time service flow measurement) 实时业务流测量
RTG (route table generator) 路由表生成器
RTMP (routing table maintenance protocol) 路由表维护协议
RTMP (real time messaging protocol) 实时消息传送协议
RTP (real time transport protocol) 实时传送协议
rtPS (real time polling service) 实时轮询服务
RTS (request to send) 请求发送
RTSE (reliable transfer service element) 可靠传输服务元素
RTSP (real time streaming protocol) 实时流媒体协议
RTT (round-trip time) 往返时间
RV (repeat visitor) 重复访客
RWA (routing and wavelength assignment) 路由和波长分配
RWR (random walk routing) 随机游走路由选择

# S

SAA（system application architecture） 系统应用体系结构

SAAL（signaling ATM adaptation layer） 信令ATM适配层

SaaS（software as a service） 软件即服务

SAC（single attachment concentrator） 单连接集中器

SACCH（slow association control channel） 慢速随路控制信道

SAE（source acceleration engine） 源站加速引擎

safety data network system (SDNS) 安全数据网络系统

safety message 安全报文

Salami technique 色拉米技术，意大利香肠术

same domain LU-LU session 同域LU-LU会话，同域逻辑单元之间会话

same frequency simulcasting network 同频同播网

same way 同路

SAML（security assertion markup language） 安全断言标记语言

SAN（storage area network） 存储区域网络

SAN（satellite access node） 卫星接入节点

SAP（service advertising protocol） 服务广告协议

SAP（service access point） 业务接入点

SAP（session announcement protocol） 会话通告协议

SAR（segmentation and reassembly） 分段和重组

SAS（single-attached station） 单连接站

SASL（simple authentication and security layer） 简单认证安全层

satellite IP broadcast network 卫星IP广播网络

SATP（simple agent transfer protocol） 简单代理传输协议

satellite access node（SAN） 卫星接入节点

satellite communication network 卫星通信网

satellite communication topology 卫星通信拓扑

satellite constellation communication network 卫星星座通信网络

satellite constellation network 卫星星座网络

satellite return Internet connection 卫星转回因特网连接

SATF（shared-access transport facility）

共享访问传输机制
saturated path 饱和路径
saturation routing 饱和路由选择
saturation signaling 饱和信令
SBL (Spamhaus block list) Spamhaus 黑名单,国际反垃圾邮件组织黑名单
SC (session control) 会话控制
SCA (security correlation agent) 安全相关代理
scaleable cluster interface (SCI) 可缩放簇接口
scaleable network 可伸缩网络
scattered-point cloud 散乱点云
scattered-point cloud data 散乱点云数据
scatternet 散射网,分布网
SCB (session control block) 会话控制块
SCCP (signaling connection control part) 信令连接控制部分
SCCPCH (secondary common control physical channel) 辅助公共控制物理信道
SCDMA (synchronous code division multiple access) 同步码分多址
SCDMA repeater SCDMA 直放站,同步码分多址直放站
SCEP (service creation environment point) 业务生成环境点
SCH (sync channel) 同步信道
SCI (scaleable cluster interface) 可缩放簇接口
SCL (synchronous connectionless link) 异步无连接
SCM (sub carrier multiplexing) 副载波复用
SCP (service control point) 业务控制点
SCR (sustainable cell rate) 可持续信元速率
screened host gateway 屏蔽主机网关
script 脚本
scripting languages 脚本语言
SCS (service capability server) 业务能力服务器
SCS (security correlation server) 安全相关服务器
SCS (structured cabling system) 结构化布线系统
SCTP (stream control transmission protocol) 流控制传输协议
SDC (synchronous directional connection) 同步定向连接
SDCCH (stand-alone dedicated control channel) 独立专用控制信道
SDH (synchronous digital hierarchy) 同步数字系列
SDH cross-connection (SDXC) SDH 交叉连接,同步数字系列交叉连接
SDH management network (SMN) SDH 管理网,同步数字系列管理网
SDI (serial digital interface) 串行数字接口
SDL (simplified data link) 简化数据链路
SDLC (synchronous data link control) 同步数据链路控制
SDLP (space data link protocol) 空间数据链路协议
SDSL (symmetrical digital subscriber

line) 对称数字用户线路
SDM (space division multiplexing) 空分复用
SDMA (space division multiple access) 空分多址
SDN (self-defending network) 自防御网络
SDN (software defined network) 软件定义网络
SDNS (safety data network system) 安全数据网络系统
SDP (session description protocol) 会话描述协议
SDR (software defined radio) 软件定义无线电
SDSL (single-line digital subscriber line) 单线数字用户线路
SDTI (serial digital transport interface) 串行数字传输接口
SDU (service data unit) 业务数据单元
SDXC (SDH cross-connection) SDH交叉连接,同步数字系列交叉连接
SE (search engine) 搜索引擎
SEAL (simple and efficient adaptation layer) 简单有效适配层
search engine (SE) 搜索引擎
search engine marketing (SEM) 搜索引擎营销
search engine optimization (SEO) 搜索引擎优化
search engine poisoning 搜索引擎投毒
search engine promotion 搜索引擎推广
search engine ranking 搜索引擎排名
search engine results page (SERP) 搜索引擎结果页面
search on Web service 基于Web服务的搜索
search Web service 搜索Web服务
secondary common control physical channel (SCCPCH) 辅助公共控制物理信道
secondary link station 次链路站
secondary logical unit (SLU) 次逻辑单元
secondary page 二级页面
secondary station 次站
second layer network loop 二层网络环路
second level domain (SLD) 二级域名
section overhead (SOH) 段开销
secure electronic transactions (SET) 安全电子交易协议
secure hypertext transfer protocol (SHTTP) 安全超文本传输协议
secure multipurpose Internet mail extensions (S/MIME) 安全多用途互联网邮件扩展
secure payment gateway 安全支付网关
secure socket layer (SSL) 安全套接层
secure shell (SSH) 安全外壳
secure socket layer virtual private network (SSL VPN) 安全套接层虚拟专用网
Security and Stable Advisory Committee (SSAC) 安全和稳定咨询委员会
security architecture 安全体系架构
security assertion markup language (SAML) 安全断言标记语言

security association (SA) 安全关联
security auditing 安全审计
security certificate 安全认证
security classification of the domain name registration system 域名注册系统安全等级
security correlation agent (SCA) 安全相关代理
security correlation server (SCS) 安全相关服务器
security hole 安全漏洞
security management (SM) 安全管理
security protocol 安全协议
security electronic payment protocol (SEPP) 安全电子付费协议
security risk assessment of information service system 信息服务业务系统安全风险评估
security risk assessment of instant messaging system 即时消息业务系统安全风险评估
security risk assessment of the domain name registration system 域名注册系统安全风险评估
security risk of information service system 信息服务业务系统安全风险
security risk of instant messaging system 即时消息业务系统安全风险
security risk of the domain name registration system 域名注册系统安全风险
security token 安全性令牌
segmentation and reassembly (SAR) 分段和重装
selective acknowledgement 选择性应答
selective band repeater 选频直放站
selective cryptographic session 选择加密会话
selective flooding routing 选择泛洪路由
selective repeat ARQ 选择性重传 ARQ
selective repeat 选择性重复
selective transmission diversity (STD) 选择发送分集
self configuration 自助配置
self-configuration network 自组织网络
self-configuration of network elements 网络元件自配置
self-defending network (SDN) 自防御网络
self healing network 自恢复网络
self healing ring (SHR) 自愈环
self-organization 自组织
self-organization network 自组织网络
self organization neural network 自组织神经网络
self-organization radio communication network 自组织无线通信网
SEM (search engine marketing) 搜索引擎营销
semantic hierarchical network model 语义层次网络模型
semantic network 语义网
semantic search 语义搜索
semantic search engine (SSE) 语义搜索引擎
semantic Web 语义网

semantic Web service search 语义网服务搜索

semantic Web stack 语义网堆栈

sender attachment delay 发送器连接延迟

sender permitted from (SPF) 发送方来源认定

sender policy framework (SPF) 发送方策略框架

send E-mail 发送电子邮件

send test E-mail 发送测试电子邮件

sending window 发送窗口

send pacing 发送调步

SEO (search engine optimization) 搜索引擎优化

separate mesh 分离网状网

separate mesh construction 分离网结构

SEPP (security electronic payment protocol) 安全电子付费协议

sequenced packet exchange (SPX) 顺序包交换

serial digital interface (SDI) 串行数字接口

serial digital transport interface (SDTI) 串行数字传输接口

serial interchange node 串行交换节点

serial line internet protocol (SLIP) 串行线路互连协议,串行链路网际协议

serial tunneling 串行隧道

serial tunneling protocol 串行隧道协议

SERP (search engine results page) 搜索引擎结果页面

server 服务器

server-based LAN 基于服务器的局域网

server-based network 基于服务器的网络

server-based spreading 基于服务器的数据发布

server clustering 服务器集群

server control 服务器控件

server farm 服务器场,服务器集群

server grabbing 服务器捕捉

server headend control 服务器前端控制

server hosting 服务器托管,主机托管

server log 服务器日志

server message block (SMB) 服务器信息块

server mirroring 服务器镜像

server rental 服务器租用,主机租用

server/requester programming interface (SRPI) 服务器/请求者编程接口

server session socket (SSS) 服务器会话套接字

server signal degrade (SSD) 服务层信号劣化

server signal fail (SSF) 服务层信号失效

server system infrastructure (SSI) 服务器系统基础结构

serveware revolution 服务件革命

service access multiplexer (SAM) 业务接入复用器

service access point (SAP) 业务接入点

service access point indicator 业务接入点标识

service adapter 服务适配器

service advertising protocol (SAP) 服务广告协议
service after sell online 在线售后服务
service capability server (SCS) 业务能力服务器
service center (SC) 服务中心
service control element 业务控制单元
service control point (SCP) 业务控制点
service creation environment point (SCEP) 业务生成环境点
service data point (SDP) 业务数据点
service data unit (SDU) 业务数据单元
service identifier (SID) 服务标识号
service independent building blocks (SIB) 业务独立构件
service level agreement (SLA) 服务等级协议
service logic (SL) 业务逻辑
service logic processing 业务逻辑处理
service logic program (SLP) 业务逻辑程序
service management access function 业务管理接入功能
service management point (SMP) 业务管理点
service network 业务网
service network of resource information 资源信息服务网络
service node (SN) 业务节点
service node interface (SNI) 业务节点接口
service of telecommunication network operation and maintenance 网络运维服务
service order table 服务次序表
service oriented architecture (SOA) 面向服务的架构
service point (SP) 业务点
service port 业务端口
service port function (SPF) 业务端口功能
service primitives 服务原语
service provider (SP) 服务提供商
service provider access 业务提供者接入
service-seeking 查找服务
service-seeking pause 查找服务暂停
service set identifier (SSID) 服务集标识
service specific connection oriented protocol (SSCOP) 特定业务面向连接协议
service specific convergence sublayer (SSCS) 特定业务汇聚子层
service specific coordination function (SSCF) 特定业务协调功能
service specific coordination function-user network interface (SSCF-UNI) 特定业务协调功能-用户网络接口
service switch access point 业务交换接入点
service switch point (SSP) 业务交换点
service traffic shaping 业务流量整形
service transparency 业务透明性
service transport node 业务传送节点
service user 服务使用者
serving GPRS supporting node (SGSN) 通用分组无线业务服务支持节点
serving radio network controller (SRNC) 服务无线网络控制器

session 会话
session announcement protocol (SAP) 会话通告协议
session awareness (SAW) data 会话内情数据
session border control 会话边界控制
session-connection synchronization 会话连接同步
session connector 会话连接器
session control (SC) 会话控制
session control block (SCB) 会话控制块
session deactivation 会话撤销
session description protocol (SDP) 会话描述协议
session hijacking 会话劫持
session information block (SIB) 会话信息块
session information retrieval (SIR) 会话信息读取
session initialization protocol (SIP) 会话初始化协议
session layer 会话层
session-level pacing 会话级调步
session limit 会话限制
session management protocol (SMP) 会话管理协议
session partner 会话伙伴
session setup failure notification (SSFN) 会话建立错误提示
session trace 会话跟踪
session trace data 会话跟踪数据
SET (secure electronic transaction) 安全电子交易协议
set asynchronous balanced mode (SABM) 设置异步平衡模式
set top box (STB) 机顶盒
SFAAC (stateful address automatic configuration) 有状态地址自动配置
SFM (source filtered multicast) 源过滤组播
SGCP (simple gateway control protocol) 简单网关控制协议
SGMP (simple gateway monitoring protocol) 简单网关监视协议
SGMP (simple gateway management protocol) 简单网关管理协议
SGSN (serving GPRS supporting node) 通用分组无线业务服务支持节点
shared-access transport facility (SATF) 共享访问传输机制
shared-control gateway 共享控制网关
shared distribution tree 共享分布树
shared information model (SIM) 共享信息模型
shared network directory 共享网络目录
shared protection ring (SPRING) 共享保护环
shared risk link group (SRLG) 共享风险链路组
shared Web hosting 网络托管共享
shared Web hosting service 网络托管共享服务
shared wireless access protocol (SWAP) 共享无线访问协议
sharing economy 分享经济,共享经济
SHDSL (single-pair high-bit-rate digital subscriber line) 单线对高位速率数字用户线路

SHDSL (symmetrical high-bit-rate digital subscriber line) 对称高位速率数字用户线路
shopping search engine 购物搜索引擎
shortest path 最短路径
shortest path algorithm 最短路径算法
shortest path first (SPF) routing 最短路径优先路由选择
shortest path tree (SPT) 最短路径树
short interframe space (SIFS) 最小帧间间隔
short message entity (SME) 短消息实体
short message gateway (SMG) 短消息网关
short message peer to peer (SMPP) 短消息点对点协议
short message service (SMS) 短消息服务
short message service point to point (SMS PP) 点对点短消息业务
short message service center (SMSC) 短信服务中心
short message service gateway 短消息网关
short message transmission protocol (SMTP) 短消息传输协议
shovelware 铲件,盗版软件
SHPW (single hop pseudo wire) 单跳伪线
SHR (self healing ring) 自愈环
SHTTP (secure hypertext transfer protocol) 安全超文本传输协议
SIB (service independent building blocks) 业务独立构件

SIB (session information block) 会话信息块
SID (service identifier) 服务标识号
SIFS (short interframe space) 最小帧间间隔
.sig 签名文件名后缀
signal ground 信号地线
signal in band 带内信号
signaling 信令
signaling ATM adaptation layer (SAAL) 信令 ATM 适配层
signaling communication network 信令通信网络
signaling connection control part (SCCP) 信令连接控制部分
signaling data link 信令数据链路
signaling data link level 信令数据链路级
signaling gateway 信令网关
signaling link (SL) 信令链路
signaling link message handler 信令链路消息处理器
signaling link test 信令链路测试
signaling link test control 信令链路测试控制
signaling link test message (SLTM) 信令链路测试消息
signaling message 信令信息
signaling message encription 信令信息加密
signaling message encription key 信令信息加密密钥
signaling message handling 信令信息处理
signaling message route 信令信息路径

signaling network 信令网
signaling network analysis system 信令网络分析系统
signaling network function 信令网络功能
signaling network management 信令网络管理
signaling network protocol 信令网络协议
signaling point (SP) 信令点
signaling point coding 信令点编码
signaling protocol 信令协议
signaling route 信令路由
signaling route management 信令路由管理
signaling route network manager (SRNM) 信令路由网络管理器
signaling route set test (SRST) 信令路由设置测试
signaling service management 信令业务管理
signaling system 信令系统
signaling system number 7 (SS7) 7号信令系统
signaling transduction network 信号传导网络
signaling transfer point (STP) 信令转接点
signaling transport protocol (SIGTRAN) 信令传输协议族
signaling virtual channel (SVC) 信令虚信道
signal out-of-band 带外信号
signal server 信号服务器
signal to interference ratio (SIR) 信号干扰比
signal-to-noise ratio (SNR) 信噪比
SIGTRAN (signaling transport protocol) 信令传输协议族
SIIT (stateless IP/ICMP translation) 无状态 IP/ICMP 翻译协议
silly-window syndrome 傻瓜窗口症状
simple agent transfer protocol (SATP) 简单代理传输协议
simple and efficient adaptation layer (SEAL) 简单有效适配层
simple authentication and security layer (SASL) 简单认证安全层
simple common gateway interface 简单通用网关接口
simple gateway 简单网关
simple gateway control protocol (SGCP) 简单网关控制协议
simple gateway management protocol (SGMP) 简单网关管理协议
simple gateway monitoring protocol (SGMP) 简单网关监视协议
simple home network management (CHNM) 简单家庭网络管理
simple mail transfer protocol (SMTP) 简单邮件传送协议
simple network management protocol (SNMP) 简单网络管理协议
simple network signaling protocol (SNSP) 简单网络信令协议
simple network time protocol (SNTP) 简单网络时间协议
simple object access protocol (SOAP) 简单对象访问协议
simple switched network 简单的交换

网路
simple wavelength allocating protocol (SWAP) 简单波长分配协议
simplex protocol for noisy channel 噪音信道单工协议
simplex stop-and-wait protocol 单工停止等待协议
simplified data link (SDL) 简化数据链路
simultaneously operating piconet (SOP) 异步操作微网
single arm routing 单臂路由
single attached station (SAS) 单连接站
single attachment concentrator (SAC) 单连接集中器
single-cable broadband LAN 单电缆宽带局域网络
single-domain network 单域网络
single fiber bi-directional 单纤双向
single hop network 单跳网络
single hop non-blocking network 单跳无阻塞网络
single hop pseudo wire (SHPW) 单跳伪线
single-line digital subscriber line (SDSL) 单线数字用户线路
single-pair high-bit-rate digital subscriber line (SHDSL) 单线对高位速率数字用户线路
single physical layer user-data switching platform architecture (SUPA) 单物理层用户数据交换平台的体系结构
single physical layer user-data switching platform architecture network (SUPANET) 单物理层用户数据交

换平台的体系结构网络
single segment pseudo wire (SSPW) 单段伪线
single sign on (SSO) 单次登录
single-station wireless network 单站无线网络
sink 信宿
sink agencies 信宿代理
sink port 信宿口
sink tree 信宿树
SIP (session initialization protocol) 会话初始化协议
SIP (SMDS interface protocol) SMDS接口协议
SIR (signal to interference ratio) 信号干扰比
SIR (session information retrieval) 会话信息读取
site 站点
site local unicast address 站点本地单播地址
site manager 站点管理员
site stickiness 网站粘性
SL (service logic) 业务逻辑
SL (signaling link) 信令链路
SLA (service level agreement) 服务等级协议
SLAAC (stateless address auto configuration) 无状态地址自动配置
SLD (second level domain) 二级域名
slice network 片式网络
sliding window 滑动窗口
sliding window flow control 滑动窗口流量控制
sliding window protocol (SWP) 滑动窗

口协议
SLIP (serial line internet protocol) 串行线路互连协议
slotted-ring control 带槽的环路控制
slotted-ring network 带槽环形网络
slot time 槽时
slow association control channel (SACCH) 慢速随路控制信道
slow start 慢启动
SLP (service logic program) 业务逻辑程序
SLTM (signaling link test message) 信令链路测试消息
SLU (secondary logical unit) 次逻辑单元
SM (security management) 安全管理
smart gateway 灵巧网关
smart hub 灵巧集线器
smart search 智能搜索
SMASH (system management architecture for server hardware) 服务器硬件的系统管理架构
SMB (server message block) 服务器信息块
SMDS (switched multi-megabit data service) 交换式多兆位数据服务
SMDS addressing SMDS 寻址,交换式多兆位数据服务寻址
SMDS customer premises equipment (CPE) SMDS 客户前端设备,交换式多兆位数据服务客户前端设备
SMDS interface protocol (SIP) SMDS 接口协议,交换式多兆位数据服务接口协议
SMDS interface protocol (SIP) levels SMDS 接口协议分级,交换式多兆位数据服务接口协议分级
SME (short message entity) 短消息实体
SMG (short message gateway) 短消息网关
smiley 表情符号,笑脸符
S/MIME (secure multipurpose Internet mail extensions) 安全多用途互联网邮件扩展
SMN (SDH management network) SDH 管理网,同步数字系列管理网
SMON (switch monitoring) 交换网络监控标准
SMP (session management protocol) 会话管理协议
SMP (service management point) 业务管理点
SMPP (short message peer to peer) 短消息点对点协议
SMS (short message service) 短信服务,短消息服务
SMSC (short message service center) 短信服务中心
SMS PP (short message service point to point) 点对点短消息业务
SMTP (simple mail transfer protocol) 简单邮件传送协议
SMTP (short message transmission protocol) 短消息传输协议
S/N (signal/noise) 信噪比
SN (service node) 业务节点
SN (super node) 超级节点
SNA (system network architecture) 系统网络体系结构

**SNALINK**（SNA network link） SNA 网络连接，系统网络体系结构网络连接

**SNA network interconnect** SNA 网络互连，系统网络体系结构网络互连

**SNA network link**（SNALINK） SNA 网络连接，系统网络体系结构网络连接

**SNA node** SNA 节点，系统网络体系结构节点

**SNAP**（subnetwork access protocol） 子网访问协议

**SNAT**（static network address translation） 静态网络地址转换

**snatch a red envelope** 抢红包

**SNC**（subnetwork connection） 子网连接

**SNDC**（subnetwork dependant convergence） 子网相关会聚层

**SNDCP**（subnetwork dependent convergence protocol） 子网相关会聚层协议

**SNI**（service node interface） 业务节点接口

**sniffer** 嗅探器

**sniffing attack** 嗅探攻击

**SNMP**（simple network management protocol） 简单网络管理协议

**SNPA**（subnetwork point of attachment） 子网连接点

**SNR**（signal-to-noise ratio） 信噪比

**SNS**（social networking service） 社交网络服务

**SNSP**（simple network signaling protocol） 简单网络信令协议

**SNTP**（simple network time protocol） 简单网络时间协议

**SOA**（service oriented architecture） 面向服务的架构

**SOAP**（simple object access protocol） 简单对象访问协议

**social engineering attack** 社会工程攻击

**social media** 社交媒体

**social media advertising** 社交媒体广告

**social media portal** 社交媒体门户

**social media network** 社交媒体网络

**social media site** 社交媒体网站

**social network** 社交网络

**social networking platform** 社交网络平台

**social networking service**（SNS） 社交网络服务

**social network site** 社交网站

**social networking system** 社交网络系统

**social networking marketing** 社交网络营销

**social software** 社会软件

**Society for Worldwide Interbank Financial Telecommunications**（SWIFT） 环球同业银行金融电信协会

**socket** 套接字

**socket address** 套接字地址

**soft error** 软错误

**soft fault** 软故障

**soft permanent virtual connection**（SPC） 软永久连接

**softswitch** 软交换

**software as a service**（SaaS） 软件即服务

software defined network (SDN) 软件定义网络
software defined radio (SDR) 软件定义无线电
software filtering 软件过滤
software radio (SWR) 软件无线电
SOH (section overhead) 段开销
SONET (synchronous optical network) 同步光纤网
songlifting 下载未授权歌曲
SOP (simultaneously operating piconet) 异步操作微网
source acceleration engine (SAE) 源站加速引擎
source address (SA) 源地址
source address routing 源地址路由选择
source address spoofing 源地址欺骗
source end station (SES) 源端站点
source-explicit forwarding 报源指定的转发
source filtered multicast (SFM) 源过滤组播
source network 报源网络
source node 源节点
source quench 报源抑制
source route 源路由
source route algorithm 源路由算法
source route bridging (SRB) 源路由桥接
source route switching 源路由交换
source route switching bridging 源路由翻译桥接
source route translation bridging 源路由翻译桥接

source route transparent bridging 源路由透明桥接
source route vector 源路由向量
source routing 源路由选择
source routing translational bridging (SR/TLB) 源路由翻译桥接
source routing transparent (SRT) 源路由透明
source service access point (SSAP) 源服务访问点
source specific multicast (SSM) 特定源组播,源特定组播,专用源组播
space communication link 空间通信链路
space data link protocol (SDLP) 空间数据链路协议
space data network 空间数据网络
space division hybrid communication 空分混合通信
space division multiple access (SDMA) 空分多址
space division multiplexing (SDM) 空分复用
space division switching 空分交换
space division switching network 空间分隔转换网络,空分交换网络
space link 空间链路
space link design algorithm 空间链路设计计算
space link identifiers 空间链路识别符
space link subnetwork 空间链路子网
space time block code (STBC) 空时分组码
space time coding (STC) 空时编码
space time transmit diversity (STTD)

空时发送分集
space time trellis code (STTC) 空时网格码
space tracking data network 空间跟踪数据网络
space tracking network 空间跟踪网
spam (stupid pointless annoying messages) 垃圾邮件
spam blacklist 黑名单
spam comments 垃圾留言,垃圾评论
spam filter 垃圾邮件过滤器
Spamhaus 国际反垃圾邮件组织
Spamhaus block list (SBL) Spamhaus 黑名单,国际反垃圾邮件组织黑名单
spam over Internet telephony (SPIT) 垃圾因特网电话
spam ticks 广告邮件的手法
spanning tree algorithm (STA) 生成树算法
spanning tree explorer (STE) 生成树探测器
spanning tree protocol (STP) 生成树协议
span of control 控制范围
span switching 区段倒换
sparse matrix canonical grid 应用稀疏矩阵规则网格
sparse grid 稀疏网格
sparse network 稀疏网络
sparse network structure 稀疏网络结构
sparse neural network 稀疏神经网络
spatial reuse protocol (SRP) 空间重用协议
spatiotemporal data 时空数据

spatiotemporal index 时空索引
SPC (soft permanent virtual connection) 软永久连接
spear phishing 鱼叉式网络钓鱼
specialized search engine (SSE) 专业搜索引擎
special net theory 狭义网论
specific application service element (SASE) 特定应用服务元素
speed-independent asynchronous sequence network 速度独立异步时序网络
SPEX (stack packet exchange) 堆栈式分组交换
SPF (sender policy framework) 发送方策略框架
SPF (sender permitted from) 发送方来源认定
SPF (service port function) 业务端口功能
SPF (shortest path first) 最短路径优先
SPIT (spam over Internet telephony) 垃圾因特网电话
SPLS (standard part library system) 标准件库系统
SPM (sync point manager) 同步点管理器
spread spectrum (SS) 扩展频谱,展频,扩频
spread spectrum communication mode 扩频通信方式
spread spectrum multipe access (SSMA) 扩频多址
SPRING (shared protection ring) 共享保护环

SPS (sync point service) 同步点服务
SPT (shortest path tree) 最短路径树
SPX (sequenced packet exchange) 顺序包交换
spyware 间谍软件
SQL injection attack SQL 注射式攻击
SRB (source route bridging) 源路由桥接
SRLG (shared risk link group) 共享风险链路组
SRNC (serving radio network controller) 服务无线网络控制器
SRNM (signaling route network manager) 信令路由网络管理器
SRPI (server/requester programming interface) 服务器/请求者编程接口
SRST (signaling route set test) 信令路由设置测试
SRT (source routing transparent) 源路由透明
SR/TLB (source routing translational bridging) 源路由翻译桥接
SRTS (synchronous residual time stamp) 同步驻留时间戳
SS (spread spectrum) 扩展频谱
SSAC (Security and Stable Advisory Committee) 安全和稳定咨询委员会
SSAP (source service access point) 源服务访问点
SSC (station selection code) 站选择代码
SSCF (service specific coordination function) 特定业务协调功能
SSCF-UNI (service specific coordination function-user network interface) 特定业务协调功能-用户网络接口
SSCOP (service specific connection oriented protocol) 特定业务面向连接协议
SSCP (system service control point) 系统服务控制点
SSCP backup 系统服务控制点后备支持
SSCP-dependent LU 依赖于系统服务控制点的逻辑单元
SSCP ID (system service control point identifier) 系统服务控制点标识符
SSCP-independent LU 独立于系统服务控制点的逻辑单元
SSCP-LU session 系统服务控制点-逻辑单元对话
SSCP monitor mode function (SMMF) 系统服务控制点监控器模式函数
SSCP-PU session 系统服务控制点-物理单元对话
SSCP rerouting 系统服务控制点重选路由
SSCP service 系统服务控制点服务程序
SSCP service manager 系统服务控制点服务管理程序
SSCP-SSCP session 系统服务控制点之间的会话
SSCS (service specific convergence sublayer) 特定业务汇聚子层
SSD (server signal degrade) 服务层信号劣化
SSE (semantic search engine) 语义搜索引擎
SSE (specialized search engine) 专业搜

索引擎
SSF（server signal fail） 服务层信号失效
SSFN（session setup failure notification） 会话建立错误提示
SSH（secure shell） 安全外壳
SSI（server system infrastructure） 服务器系统基础结构
SSI（switch-to-switch interface） 交换机到交换机接口
SSID（service set identifier） 服务集标识
SSL（secure socket layer） 安全套接层
SSL Portal VPN 安全套接层门户虚拟专用网
SSL Tunnel VPN 安全套接层隧道虚拟专用网
SSM（source specific multicast） 特定源组播
SSMA（spread spectrum multipe access） 扩频多址
SSP（service switch point） 业务交换点
SSPW（single segment pseudo wire） 单段伪线
SSS（server session socket） 服务器会话套接字
SS7（signaling system number 7） 7号信令系统
ST（stream protocol） 流协议
STA（spanning tree algorithm） 生成树算法
stackable hub 可堆叠集线器
stackable switch 可堆叠交换机
stack packet exchange（SPEX） 堆栈式分组交换

stand-alone cloud 单云，独立云
stand-alone dedicated control channel（SDCCH） 独立专用控制信道
stand-alone network system 独立网络系统
standard Ethernet 标准以太网
standard generalized markup language（SGML） 标准通用标记语言
standard part library system（SPLS） 标准件库系统
star network 星形网络，星状网
star network topology structure 星型网络拓扑结构
star/ring network 星形/环形网络
start of text（STX） 正文开始
star topology 星形拓扑
state-based application layer protocol 基于状态的应用层协议
stateful address automatic configuration（SFAAC） 有状态地址自动配置
stateful files server 全状态文件服务器
stateless address auto configuration（SLAAC） 无状态地址自动配置
stateless and call-back filing system 无状态和回调文件系统
stateless files service 无状态文件服务器
stateless IP/ICMP translation（SIIT） 无状态 IP/ICMP 翻译协议
stateless protocol 无状态协议
statically defined peer-to-peer session 静态定义的平等会话
statistical time-division multiplexing（STDM） 统计时分多路复用

static and dynamic interior routes 静态与动态内部路由
static default route 静态默认路由
static domain name resolution 静态域名解析
static firewalls 静态防火墙
static integrated network access 静态集成网络访问
static network address translation (SNAT) 静态网络地址转换
static routing 静态路由选择
static routing and wavelength assignment 静态路由和波长分配
static Web page 静态网页
station management (SMT) 站管理
STB (set top box) 机顶盒
STBC (space time block code) 空时分组码
STC (space time coding) 空时编码
STD (selective transmission diversity) 选择发送分集
STDM (synchronous time division multiplexer) 同步时分复用器
STDM (statistical time-division multiplexing) 统计时分多路复用
STE (spanning tree explorer) 生成树探测器
sticker gallery 表情包
sticky 帖子置顶
STM-N (synchronous transfer module-N) N 阶同步传输模式
storage area network (SAN) 存储区域网络
storage management interface specification (SMI-S) 存储管理接口规范

storage medium 存储媒体
Storage Networking Industry Association (SNIA) 网络存储工业协会
storage virtualization 存储虚拟化
store and forward 存储转发
store and forward mode 存储转发模式
store and forward network 存储转发网络
store and forward switching 存储转发交换技术
store and forward switching center 存储转发交换中心
store and forward system 存储转发系统
STP (signaling transfer point) 信令传输点
STP (spanning tree protocol) 生成树协议
stream cipher 流密码
streaming content delivery network 流媒体内容分发网络
streaming distribution network 流(媒体)分发网络
streaming media 流媒体
streaming media network 流媒体网络
streaming network 流式网络
streaming technology 流式技术
stream media delivery network 流媒体分发网络
stream-oriented 面向流的
stream protocol (ST) 流协议
stream socket 流式套接字
strict non-blocking network 严格无阻塞网络
strongly connected random network 强

连通随机网络
structured cabling system (SCS) 结构化布线系统
structured data transfer (SDT) 结构数据传输
STTC (space time trellis code) 空时网格码
STTD (space time transmit diversity) 空时发送分集
stub network 存根网络
stupid pointless annoying messages (spam) 垃圾邮件
subarea 子区域
subarea address 子区域地址
subarea host node 子区域宿主节点
subarea link 子区域链路
subarea logical unit 子区域逻辑单元
subarea network 子区域网络
subarea node 子区域节点
subarea path control 子区域路径控制
subarea physical unit 子区域物理单元
subarea routing function 子区域路由函数
subdomain 子域名
sublayer 子层
submit post 提交贴子
subnet 子网
subnet accessing 子网接入
subnet addressing 子网编址
subnet hiding 子网屏蔽
subnet isolation 子网隔离
subnet mask 子网掩码
subnet prefix 子网前缀
subnet system 子网系统
subnetwork access protocol (SNAP) 子网访问协议
subnetwork connection (SNC) 子网连接
subnetwork dependant convergence (SNDC) 子网相关会聚层
subnetwork dependent convergence protocol (SNDCP) 子网相关会聚层协议
subnetwork layer 子网层
subnetwork point of attachment (SNPA) 子网连接点
subnetwork protection 子网保护
subnetwork space 子网络空间
sub-optimum routing algorithm 次优路由算法
subordinate page 二级页面
summary address 概括地址
SUPA (single physical layer user-data switching platform architecture) 单物理层用户数据交换平台的体系结构
SUPANET (single physical layer user-data switching platform architecture network) 单物理层用户数据交换平台的体系结构网络
super cluster center node 超级簇中心节点
super cluster control station 超级集群控制站
super cluster Web server 集群超级网络服务器
supernet 超网
supernet block 超网块
supernet mask 超网掩码
supernet number 超网号
super node (SN) 超级节点

super node selection 超级节点选取
super node P2P network 超级节点对等网络,超级节点 P2P 网络
superserver 超级服务器
supervisory channel 管理信道
supplicant 申请者
support network 支撑网
surf the net 上网,网上冲浪
survivability network systems 可生存网络系统
sustainable cell rate (SCR) 可持续信元速率
sustained packet rate 持续分组速率
SVC (switched virtual circuit) 交换虚电路
SVC (switched virtual connection) 交换虚拟连接
SVC (signalling virtual channel) 信令虚信道
SVCC (switched virtual channel connection) 交换的虚拟通道连接
SVPC (switched virtual path connection) 交换的虚拟路径连接
SWAP (shared wireless access protocol) 共享无线访问协议
SWAP (simple wavelength allocating protocol) 简单波长分配协议
swapper 换客
sweep-code payment 扫码支付
SWIFT (Society for Worldwide Interbank Financial Telecommunications) 环球同业银行金融电信协会
SWIFT code 环球同业银行金融电信协会代码
switched intermediate network node 开关中间网络节点
switched LAN 交换式局域网
switched multi-megabit data service (SMDS) 交换式多兆位数据服务
switched network 交换网络
switched network backup 交换网络备份
switched network block 交换网络功能块
switched point-to-point topology 交换式点到点拓扑结构
switched service network 交换服务网
switched SNA major node 交换式系统网络体系结构主节点
switched telecommunications network 远程通信交换网络
switched telephone network 电话交换网络
switched virtual channel connection (SVCC) 交换的虚拟通道连接
switched virtual circuit (SVC) 交换虚电路
switched virtual connection (SVC) 交换虚拟连接
switched virtual path connection (SVPC) 交换的虚拟路径连接
switches loop 交换机环路
switches loop monitoring 交换机环路监测
switch everywhere 全部用交换
switching hub 交换式集线器,交换器
switching network capacity 交换网络容量
switch monitoring (SMON) 交换网络监控标准

SWP (sliding window protocol) 滑动窗口协议
SWR (software radio) 软件无线电
symmetrical digital subscriber line (SDSL) 对称数字用户线路
symmetrical high-bit-rate digital subscriber line (SHDSL) 对称高位速率数字用户线路
symmetric network address translation 均衡网络地址转换
sync channel (SCH) 同步信道
synchronization signal unit (SYU) 同步信号单元
synchronized multimedia integration language (SMIL) 同步多媒体集成语言
synchronizing stream cipher 同步流密码
synchronous code division multiple access (SCDMA) 同步码分多址
synchronous connectionless link (SCL) 异步无连接
synchronous data link control (SDLC) 同步数据链路控制
synchronous data network 同步数据网络
synchronous data transfer 同步数据传送
synchronous digital hierarchy (SDH) 同步数字系列
synchronous directional connection (SDC) 同步定向连接
synchronous optical network (SONET) 同步光纤网
synchronous protocol 同步协议
synchronous residual time stamp (SRTS) 同步驻留时间戳
synchronous serial transmission 同步串行传输
synchronous time division multiplexer (STDM) 同步时分复用器
synchronous transfer mode (STM) 同步传输模式
synchronous transfer mode network 同步传输模式网
synchronous transfer module-N (STM-N) N 阶同步传输模式
synchronous transport lane (STL) 同步传送通道
synchronous transport module (STM) 同步传送模块
sync point manager (SPM) 同步点管理器
sync point service (SPS) 同步点服务
SYN flood attack 同步泛洪攻击,SYN 泛洪攻击
syndrome of network addiction 网络成瘾综合症
system application architecture (SAA) 系统应用体系结构
system management architecture for server hardware (SMASH) 服务器硬件的系统管理架构
system network architecture (SNA) 系统网络体系结构
system service control point (SSCP) 系统服务控制点
system service control point identifier (SSCP ID) 系统服务控制点标识符
system slowdown 系统减速

**system management architecture for server hardware (SMASH)** 服务器硬件的系统管理架构

**system supervisor** 系统管理程序

**SYU (synchronization signal unit)** 同步信号单元

# T

T (tera) 垓,太
TA (target audience) 目标受众
TA (timing advance) 定时提前
table-driven routing 表驱动路由
TACACS (terminal access contorller access control system) 终端访问控制器访问控制系统
tagged cell rate (TCR) 标志的信元速率
tag switching 标记交换
tamed frequency modulation (TFM) 平滑调频
target audience (TA) 目标受众
target token-rotation time (TTRT) 目标令牌循环时间
TB (terabyte) 兆兆字节,太字节,万亿字节
TB (transport block) 传输块
TBS (transport block set) 传输块集
TC (transmission convergence) 传输会聚
TCAM (tele communication access method) 远程通信访问法
TCM (time compression multiplexing) 时间压缩复用
TCO (total cost of ownership) 总体拥有成本
TCP (transmission control protocol) 传输控制协议
TCP (trusted computing platform) 可信计算平台
TCP/IP (transmission control protocol/Internet protocol) 传输控制协议/网际协议
TCR (tagged cell rate) 标志的信元速率
TCS (transmission convergence sublayer) 传输会聚子层
TCU (trunk coupling unit) 干线耦合器
TDD (time division duplex) 时分双工
TDI (transport driver interface) 传输驱动接口
TDM (time-division multiplexing) 时分复用
TDMA (time division multiple access) 时分多址
TDM over GEM GEM 帧承载 TDM 业务
TDNW (time division network) 时分网络
TDR (time domain reflection) 时域反射
TD-SCDMA (time division-synchronous code division multiple access) 时分同步码分多址

technical and office protocol (TOP) 技术与办公协议
technical architecture group (TAG) 技术构架组
teleaction service 遥信业务
telebanking 远程银行
telecom carrier class IP network 电信级 IP 网
telecom integrated network management 电信综合网管
telecom management network 电信管理网
telecommunication access method (TCAM) 远程通信访问法
Telecommunication Industries Association (TIA) （美国）电信工业协会
telecommunication management network (TMN) 电信管理网
telecommunication network 电信网
telecommunication network maintenance 电信网络维护
telecommunication networks security requirements 电信网络安全要求
telecommunication service network 电信业务网
telecommunication supporting network 电信支撑网
telecommunication system 电信系统
telecommuter 远程办公者
telecom network fraud 电信网络诈骗
telecom network system 电信网络系统
telecom service network 电信业务网
teleconferencing 远程会议
Telemanagement Forum (TMF) 电信管理论坛
TELEMED (telemedicine system) 远程医疗系统
telemedicine system (TELEMED) 远程医疗系统
telemetry asynchronous block serial protocol (TABS) 遥测异步块串行协议
telnet internal network protocol (TINP) 电信网内部网络协议
telnet processor reporting facility (TPRF) 电信网处理器报告工具
teleordering 远程订货
telephone bank service (TBS) 电话银行服务
telephone number mapping (ENUM) 电话号码映射
teleprocess 远程处理
teleputer 电视计算机
telepresence 远程呈现，远程监控
teleshopping 远程购物
teleworking 远程办公
Telnet 远程登录
temporal key integrity protocol (TKIP) 临时密钥完整性协议
temporary block flow (TBF) 临时数据块流
temporary connection 暂时连接
temporary internet files (TIF) 临时网络文件夹
temporary link layer identity 临时链路层识别
temporary link level identity (TLLI) 临时链路级识别
temporary logical link identification (TLLI) 临时逻辑链路标识

temporary redirect 暂时重定向
temporary trunk blocking (TTB) 中继线暂时拥塞
ten gigabit Ethernet 万兆以太网
terabyte (TB) 太字节,兆兆字节
terminal access contorller access control system (TACACS) 终端访问控制器访问控制系统
terminal access controller (TAC) 终端访问控制器
terminal adapter (TA) 终端适配器
terminal equipment subport (TESP) 终端设备子端口
terminal equipment type 1 (TE1) 一类终端设备,标准 ISDN(综合业务数字网)终端
terminal equipment type 2 (TE2) 二类终端设备,非标准 ISDN(综合业务数字网)终端
terminal interface processor (TIP) 终端接口处理机
terminal/modem interfaces 终端/调制解调器接口
terminal network 终端网络
terminal of a network 网络的端
terminal-oriented network 面向终端网络
terminal pair 端对
terminal port 终端端口
terminal power over Ethernet (TPOE) 以太网反向馈电
terminal protocols 终端协议
terminal server 终端服务器
terminal session 终端会话时间[通话期]

terminating network 终接网
terminator cap 端接器
text processing network 文本处理网络
TFC (transport format combination) 传输格式组合
TFCI (transport format combination indicator) 传输格式组合指示
TFCS (transport format combination set) 传输格式组合集
TFI (transport format indicator) 传输格式指示
TFS (transport format set) 传输格式集
TFRC (transmission format resource composite) 传输格式和资源组合
TFTP (trivial file transfer protocol) 简单文件传输协议
TGDK (transparent gateway developers kit) 透明网关开发工具箱
TGMS (third generation mobile system) 第三代移动通信系统
TH (time hopping) 跳时
TH (transimssion header) 传输标题
thanks in advance (TIA) 提前感谢
the No. 7 signaling network 七号信令网络
thick Ethernet 粗缆以太网
thicknet 粗网
thin client 瘦客户(机)
thin client computer 瘦客户机电脑
thin client computing 瘦客户机计算
thin client/fat server 瘦客户/胖服务器
thin client mode 瘦客户机模式
thin client/server 瘦客户机/服务器

thin Ethernet 细缆以太网
thinnet 细网
thinnet cable 细缆
thinnet coax 同轴细缆
third generation mobile communication technology (3G) 第三代移动通信技术
third generation mobile system (TGMS, 3G) 第三代移动通信系统
third generation partnership project (3GPP) 第三代移动通信项目组织
third generation partnership project 2 (3GPP2) 第三代移动通信项目组织2
third generation wireless (3G) 第三代无线技术
three layers network framework 三层网络架构
three layers network loop 三层网络环路
third level domain 三级域名
third party call control (TPCC) 第三方调用协议
third-party E-commerce payments 电子商务第三方支付
third-party E-commerce platform 第三方电子商务平台
third party electronic payments 第三方电子支付
third party payment gateway 第三方支付网关
third party payment of Internet 网络第三方支付
third party service provider 第三方服务提供商

three network in one network fusion 三网融合
tiled convolutional neural network 平铺卷积神经网络
TLLI (temporary logical link identification) 临时逻辑链路标识
TLP (transport layer protocol) 传输层协议
TLS (transparent LAN service) 透明局域网业务
TLS (transport layer security) 传输层安全
thread 跟帖
three gold engineerings "三金"工程
three layers of Internet service 因特网服务的三个层次
three phases of microcomputer networking 微机组网的三个阶段
three phases of X.25 connection X.25连接建立的三阶段
three primary classes of IP addresses 三类主要IP地址
three tier architecture (TTA) 三层体系结构
three-tier client/server 三层客户机/服务器(结构)
three-way-handshake 三段[次]握手
threshold window 阈值窗口,门限窗口
throughput class 吞吐量等级
throughput class negotiation 吞吐量等级协商
throughput negotiation 吞吐量协商
THSS (time hopping spread spectrum) 跳时扩频
THT (token-holding time) 令牌保持

时间
TIA (Telecommunication Industries Association) （美国）电信工业协会
TIA (thanks in advance) 先表谢意
TIC (token-ring interface coupler) 令牌环网接口耦合器
TIF (temporary internet files) 临时网络文件夹
time and date service 时间和日期服务
time-based connection management 基于计时器的连接管理
time compression multiplexing (TCM) 时间压缩复用
time-delay and data-packet dropout 时延与数据包丢失
time-division channel 时分信道
time-division code division multiple access (TD-CDMA) 时分码分多址
time division duplex (TDD) 时分双工
time division hybrid communication 时分混合通信
time division multiple access (TDMA) 时分多址,时分多路访问
time-division multiplexed switching (TMS) 时分复用交换
time division multiplexing (TDM) 时分复用
time division network (TDNW) 时分网络
time division switching 时分交换
time division-synchronous code division multiple access (TD-SCDMA) 时分同步码分多址
time-division transfer 分时传递
time-division tube 时间分割管

time domain method 时域理论
time domain cluster 时域簇
time domain reflection (TDR) 时域反射
time domain sampling 时域采样
time domain sensitivity 时域灵敏度
time of arrival (ToA) 到达时间
time out factor (TOF) 超时因子
time server 时间服务器
time-sharing polling 分时轮询
time slot (TS) 时隙
time slot controlling 时隙控制
time slot exchanger 时隙交换器
time slot interchange (TSI) 时隙交换
time slot mismatching 时隙错连
time slot ring 时隙环
time slot sequence integrity (TSSI) 时隙序列完整性
time slot switching 时隙交换
time source service 时间源服务
time stamp 时标,时间戳
time stamp authority 时间戳服务中心
time stamp counter (TSC) 时间戳计数器
time stamp protocols 时间戳协议
time stamp recognition 时间戳识别
time stamp reply 时间戳应答
time stamp request and reply 时间戳请求和应答
time stamp system 时间戳系统
time strategy slot 时间策略槽
time synchronization service 时间同步服务
time to live (TTL) 生存时间
timing advance (TA) 定时提前

TINP (telenet internal network protocol) 电信网内部网络协议
TIP (terminal interface processor) 终端接口处理机
T-interface T接口
TLI (transport level interface) 传输层界面
TLLI (temporary logical link identification) 临时逻辑链路标识
TLP (transmission level point) 传输电平点
TLS (transport layer security) 传输层安全
TLS (transparent LAN service) 透明局域网业务
TLSP (transportlayer security protocol) 传输层安全协议
TMN (tele communication management network) 电信管理网
TMPLS (transport multiprotocol label switching) 传输层多协议标记交换
TMS (time-division multiplexed switching) 时分复用交换
TMux (transport multiplexing protocol) 传输多路复用协议
TNA (trusted network architecture) 可信网络架构
TNC (trusted network connection) 可信网络连接
TNCS (trusted network connect specifications) 可信网络连接规范
TNDU (transport network data unit) 传输网络数据单元
TNIU (trustworthy network interface unit) 可信网络接口单元

TNP (trusted network platform) 可信网络平台
ToA (time of arrival) 到达时间
TOF (time out factor) 超时因子
token 令牌
token access control 令牌存取控制
token access protocol 令牌访问协议
token bucket 令牌桶
token bus 令牌总线
token bus network 令牌总线网
token-holding time (THT) 令牌保持时间
token leaky bucket algorithm 令牌漏桶算法
token management 令牌管理
token passing 令牌传递
token passing bus 令牌传递总线
token passing control protocol 令牌传递控制协议
token passing network 令牌传递网
token passing policy 令牌传递策略
token passing procedure 令牌传递规程
token passing ring 令牌传递环
token passing ring network 令牌传递环网络
token passing system 令牌传送系统
token ring 令牌环
token-ring adapter type 1 令牌环网适配器类型1
token-ring adapter type 2 令牌环网适配器类型2
token-ring interface coupler (TIC) 令牌环网接口耦合器
token ring LAN 令牌环局域网

token-ring MAC sublayer protocol　令牌环介质访问控制子层协议
token-ring maintenance　令牌环维护
token ring mechanism　令牌环机制
token-ring network　令牌环网
token-ring network types　令牌环形网类别
token-ring policy　令牌环策略
TOP (technical and office protocol)　技术与办公协议
top level domain (TLD)　顶级域名
topology and routing service (TRS)　拓扑和路由服务
topology database　拓扑数据库
topology database update (TDU)　拓扑数据库更新
topology of network　网络拓扑
TOS (type of service)　服务类型
total area network　总域网
total cost of ownership (TCO)　总体拥有成本
TPC (transmit power control)　发射功率控制
TPCC (Transaction Processing Performance Council)　事务处理性能委员会
TPCC (third party call control)　第三方调用协议
TPD (trailing packet discard)　尾帧丢弃
TP-DDI (twisted-pair distributed data interface)　双绞线分布式数据接口
TPDU (transport protocol data unit)　传输协议数据单元
TP-MIC (twisted-pair media interface connector)　双绞线媒体接口连接器
TPOE (terminal power over Ethernet)　以太网反向馈电
TPRF (telenet processor reporting facility)　电信网处理器报告工具
Tracert (trace router)　跟踪路由
tracking website hits　追踪网站点击数
traffic contract　业务量合约
traffice class　业务类别
traffic engineering (TE)　流量工程
traffic exchange alliance　流量交换联盟
traffic flooding attack　大流量洪泛攻击
traffic management　交通管理
traffic adaptive medium access　流量自适应介质访问
traffic requirement matrix　流通量要求矩阵
traffic shaping (TS)　流量整形
trailer　报尾
trailer protocol　报尾协议
trailing packet discard (TPD)　尾帧丢弃
trail signal degrade (TSD)　路径信号劣化
trail signal fail (TSF)　路径信号失效
transaction-based routing　基于事务的路由选择
transaction capabilities (TC)　事务处理能力
transaction capabilities applications part (TCAP)　事务能力应用部件
Transaction Processing Performance Council (TPCC)　事务处理性能委

员会
transaction service layer　事务服务层
transaction tracking system（TTS）　事务跟踪系统
transactive content　可交易内容
transfer failure probability　传递失败概率
transient buffer exposure　暂态缓存量
transit delay　传输延迟
transit delay selection and indication（TDSAI）　转接时延选择和指示
transitional pages　过渡页
transit network　转接网
transit network identification　转接网络标识
translation bridging　翻译桥接
transmission access network　传输接入网
transmission control characters（TCC）传输控制字符
transmission control layer　传输控制层
transmission control protocol（TCP）传输控制协议
transmission control protocol/internet protocol（TCP/IP）传输控制协议/网际协议
transmission convergence（TC）传输会聚
transmission convergence sublayer（TCS）传输会聚子层
transmission format resource composite（TFRC）传输格式和资源组合
transmission frame　传输帧
transmission group（TG）传输组
transmission group identifier（TGID）

传输组标识符
transmission header（TH）传输标题
transmission limit　传输极限
transmission management network　传送管理网
transmission medium　传输媒体
transmission mode　传输方式
transmission network　传输网
transmission network integrated management system　传输网综合管理系统
transmission network management agent　传输网管代理器
transmission network management system　传输网管理系统
transmission path delay　传输路径延迟
transmission path level　传输通路级
transmission path loss　传输路径损耗
transmission priority　传输优先级
transmission protocol　传输协议
transmission service profile　传输服务轮廓文件
transmission subsystem　传输子系统
transmission time interval（TTI）传输时间间隔
transmit burst　传输突发
transmit control block　传输控制块
transmit copy line　发送复制线
transmit network　传输网络
transmit pair　传送线对
transmit port　传输端口
transmit power control（TPC）发射功率控制
transparent bridge　透明式网桥,透明式桥接器

transparent bridge algorithm 透明网桥算法

transparent bridging 透明桥接

transparent code data transmission 透明代码数据传输

transparent communication 透明通信

transparent data 透明数据

transparent data communication code 透明数据通信码

transparent data link 透明数据链路

transparent data migration 透明数据迁移

transparent data reduction 透明数据压缩

transparent data transfer module 透明数传模块

transparent encryption mode 透明加密模式

transparent fragmentation 透明报片分割

transparent gateway 透明网关

transparent gateway developers kit (TGDK) 透明网关开发工具箱

transparent grid 透明网格

transparent intelligent network 透明智能网络

transparent LAN 透明局域网

transparent LAN service (TLS) 透明局域网业务

transparent mobile IP technology 透明移动IP技术

transparent mode 透明模式

transparent network 透明网络

transparent network security 透明网络安全

transparent network services 透明的网络服务

transparent network substrate 透明网络底层

transparent optical network 透明光网络

transparent subnet 透明子网络

transparent task to task communication 透明任务到任务的通信

transparent text transmission 透通本文传输

transparent transmission 透明传输

transparent video gateway 透明视频网关

transport block (TB) 传输块

transport block set (TBS) 传输块集

transport control protocol (TCP) 传输控制协议

transport driver interface (TDI) 传输驱动接口

transport format combination (TFC) 传输格式组合

transport format combination indicator (TFCI) 传输格式组合指示

transport format combination set (TFCS) 传输格式组合集

transport format indicator (TFI) 传输格式指示

transport format set (TFS) 传输格式集

transport function (TF) 传送功能

transport layer 传输层

transport layer protocol (TLP) 传输层协议

transport layer security (TLS) 传输层

安全

transport layer security protocol (TLSP) 传输层安全协议

transport level interface (TLI) 传输层界面

transport multiplexing protocol 传输多路复用协议

transport multiprotocol label switching (TMPLS) 传输层多协议标记交换

transport network 传送网

transport network data unit (TNDU) 传输网络数据单元

transport protocol data unit (TPDU) 传输协议数据单元

transport protocol for real-time application 实时应用传送协议

transport protocol identification mechanism 传输协议识别机制

transport protocol machine 传输协议机

transport service access pointer (TSAP) 传输服务访问点

transport service data unit (TSDN) 传输服务数据单元

tree topology 树型拓扑

tributary unit group (TUG) 支路单元组

tri-network convergence (TNC) 三网融合

triplexer optical assembly 三端口光电组件

trivial file transfer protocol (TFTP) 简单文件传输协议

TRS (topology and routing service) 拓扑和路由服务

trunk coupling unit (TCU) 干线耦合器

trunk media gateway 中继媒体网关

trusted control network 可信控制网络

trusted identification 前向可信验证

trusted network 可信网络

trusted network access 可信网络接入

trusted network architecture (TNA) 可信网络架构

trusted network connection (TNC) 可信网络连接

trusted network connect specifications (TNCS) 可信网络连接规范

trusted network connect subgroup 可信网络连接小组

trusted network operation 网络可信运行

trustworthy network interface unit 可信网络接口单元

trusted network platform (TNP) 可信网络平台

trusted third party (TTP) 可信第三方

trusted time stamp 可信时间戳

trusted transaction 可信交易

trust relationship 委托关系

trust service 信任服务

trustworthy network interface unit (TNIU) 可信网络接口单元

trustworthy terminal interface unit (TTIU) 可信终端接口单元

TS (time slot) 时隙

TS (transmission service) 传输服务

TS (traffic shaping) 流量整形

TSC (time stamp counter) 时间戳计数器

TSD (trail signal degrade) 路径信号劣化
TSDU (transport service data unit) 传输服务数据单元
TSF (trail signal fail) 路径信号失效
TTA (three tier architecture) 三层体系结构
TTB (temporary trunk blocking) 中继线暂时拥塞
TTI (transmission time interval) 传输时间间隔
TTIU (trustworthy terminal interface unit) 可信终端接口单元
TTL (time to live) 生存时间
TTLS (tunneled transport layer security) 隧道传输层安全
TTP (trusted third party) 可信第三方
TTRT (target token-rotation time) 目标令牌循环时间
TUG (tributary unit group) 支路单元组
tunnel broker 隧道代理
tunneled transport layer security (TTLS) 隧道传输层安全
tunneling protocol (TP) 隧道协议
twisted-pair distributed data interface (TP-DDI) 双绞线分布式数据接口
twisted-pair media interface connector (TP-MIC) 双绞线媒体接口连接器
Twitter 推特
two important boundaries in TCP/IP model TCP/IP模型两个重要边界
two stage oscillation 双阶段振荡
two-tier client/server 两层客户机/服务器
two-tiered Internet 两层因特网
two-way transparent transmission 双向透明传输
type-II hybrid automatic repeat request (HARQ) 第二类混合自动重发请求
type of service (TOS) 服务类型
type of service routing 服务类型路由选择
TYVM (thank you very much) 非常谢谢

# U

UADSL(usual asymmetric digital subscriber line) 通用非对称数字用户线路

UAE(user acceleration engine) 用户加速引擎

UAL(user access line) 用户接入线路

UAN(user access network) 用户接入网

UATI(unicast access termination identifier) 单点广播接入终端标识

UAWG(universal asymmetric digital subscriber line working group) 通用非对称数字用户线工作组

UBE(unsolicited bulk email) 未经许可的大宗邮件

ubiquitous convergent network 泛在融合网络

UCE(unsolicited commercial email) 未经许可的商业邮件

ubiversal computing network 泛在计算网络

ubiquitous computing system 泛在计算系统

ubiquitous network 泛在网,泛在网络

ubiquitous sensor network(USN) 泛在传感器网络

UBR(unspecified bit rate) 未指定比特率

UCAID(University Corporation for Advanced Internet Development) 先进因特网技术开发大学联盟

UDDI(universal description, discovery, and integration) 通用描述、发现与集成

UDP(user datagram protocol) 用户数据报协议

UDR(user destination routing) 用户目标路由选择

UE(user equipment) 用户设备

UE(user element) 用户元素

UE(user experience) 用户体验

UED(user experience design) 用户体验设计

UGC(user generated content) 用户生成内容

UGS(unsolicited grant service) 主动授予服务

UIC(user identification code) 用户标识码

UIM(user identity model) 用户识别模块

UIN(universal Internet number) 通用因特网号码

UIP(user interface program) 用户接口程序

UL(uplink) 上行链路

ULA (unique local IPv6 unicast address) 唯一本地 IPv6 单播地址
ULHOT (ultra long haul optical transmission) 超长光传输
ULP (upper-layer protocol) 较高层协议
ultimate destination 最终报宿
ultra mobile broadband (UMB) 超移动宽带
ultra wideband (UWB) 超宽带
UMA (unlicensed mobile access) 非授权移动接入
UMB (ultra mobile broadband) 超移动宽带
UME (user network interface management entity) 用户网络接口管理实体
UMS (unified messaging system) 统一消息系统
UMTS (universal mobile telecommunication system) 通用移动通信系统
UMTS terrestrial radio access network (UTRAN) UMTS 陆地无线接入网,通用移动通信系统陆地无线接入网
UNA (upstream neighbor's address) 上游相邻地址
unassigned cells 未赋值信元
unauthorized APPN end node 未授权 APPN 终点节点
UNC (universal naming convention) 通用命名约定
unfollow 取消关注
UNI (user network interface) 用户网络接口
unicast 单播

unicast access 单播访问
unicast access termination 单点广播接入终端
unicast access termination identifier (UATI) 单点广播接入终端标识
unicast address 单播地址
unicast reverse path forward (URPF) 单播逆向路径转发
unicast transmission 单播传输
unidirectional path protected ring (UPPR) 单向通道保护环
unidirectional ring 单向环
uniform domain name dispute resolution mechanism 统一域名争议解决机制
unified messaging system (UMS) 统一消息系统
uniform resource indentifier (URI) 统一资源标识符
uniform resource locator (URL) 统一资源定位符
uniform resource name (URN) 统一资源名
unifying slot assignment protocol (USAP) 统一时隙分配协议
unilateral path switching ring (UPSR) 单向同步系统
UnionPay 银联
UnionPay card 银联卡
UnionPay QuickPass 银联闪付
unique local IPv6 unicast address (ULA) 唯一本地 IPv6 单播地址
unique IP (UI) 独立 IP
unique visitor (UV) 独立访客
unique select slave 唯一选择从属
united threat management (UTM) 统

—威胁管理
unity gain  整体增益
universal address administration  全局地址管理
universal asymmetrical digital subscriber line (UADSL)  通用非对称数字用户线路
universal asymmetric digital subscriber line working group (UAWG)  通用非对称数字用户线工作组
universal description, discovery, and integration (UDDI)  通用描述、发现与集成
universal Internet number (UIN)  通用因特网号码
universally administered address  通用监管地址
universal mobile telecommunication system (UMTS)  通用移动通信系统
universal naming convention (UNC)  通用命名约定
universal networking language (UNL)  通用网络语言
universal protocol platform (UPP)  通用协议平台
universal resource locator (URL)  通用资源定位器
universal serial bus (USB)  通用串行总线
universal SIM application tool (USAT)  通用SIM应用工具包
universal telecommunication radio access network (UTRAN)  通用电信无线接入网
universal test and operations interface for ATM  ATM通用测试和操作接口
universal unique identifier (UUID)  通用唯一标识符
University Corporation for Advanced Internet Development (UCAID)  先进因特网技术开发大学联盟
Unix-to-Unix copy program (UUCP)  Unix到Unix复制程序
UNL (universal networking language)  通用网络语言
unlicensed mobile access (UMA)  非授权移动接入
unmoderated mailing list  非仲裁邮递表
unrestricted simplex protocol  无约束单工协议
unshielded twisted pair (UTP)  非屏蔽双绞线
unsolicited bulk email (UBE)  未经许可的大宗邮件
unsolicited commercial email (UCE)  未经许可的商业邮件
unsolicited grant service (UGS)  主动授予服务,非申请授予业务
unspecified address  未指定地址
unspecified bit rate (UBR)  未指定位速率
unstructured supplementary service data (USSD)  非结构化的补充数据业务
UpA (Uploading Accelerator)  上传加速
uplink (UL)  上行链路
uplink shared channel (USCH)  上行共享信道
uplink state flag (USF)  上行链路状态

标记

Uploading Accelerator（UpA） 上传加速

UPMA（user-dependent perfect-scheduling multiple access） 基于用户的完美调度多路访问

UPP（universal protocol platform） 通用协议平台

up path 上行路径

upper-layer protocol（ULP） 较高层协议

upper-layer protocol drive 上层协议驱动

upper-layers 高层

UPPR（unidirectional path protected ring） 单向通道保护环

UpPTS（uplink pilot time slot） 上行导频时隙

UPSR（unilateral path switching ring） 单向同步系统

upstream bandwidth 上行带宽

upstream burst communication 上行突发通信

upstream line 上游线路

upstream neighbor's address（UNA） 上游相邻地址

upward communication 向上通信

upward-multiplexing 向上多路转接

urgent arbitration 紧急仲裁

urgent data 紧急数据

URI（uniform resource identifier） 统一资源标识符

URL（uniform resource locator） 统一资源定位符

URL depth URL 深度，统一资源定位符

URL poisoning URL 投毒，统一资源定位符

URL server type URL 服务器类型，统一资源定位符

URN（uniform resource name） 统一资源名

urn protocol 壶球协议

URPF（unicast reverse path forward） 单播逆向路径转发

usage parameter control（UPC） 使用参数控制

USAP（unifying slot assignment protocol） 统一时隙分配协议

USAT（universal SIM application tool） 通用 SIM 应用工具包

USB（universal serial bus） 通用串行总线

USB transport mode USB 传输方式

USCH（uplink shared channel） 上行共享信道

Usenet news Usenet 新闻

Usenet newsgroup Usenet 新闻组

user acceleration engine（UAE） 用户加速引擎

user access line（UAL） 用户接入线路

user access network（UAN） 用户接入网

user agent（UA） 用户代理

user application network 用户应用网络

user class of service 用户服务级别

user datagram protocol（UDP） 用户数据报协议

user-dependent perfect-scheduling multiple

access protocol　基于用户的完美调度多路访问协议

user-dependent perfect-scheduling multiple access (UPMA)　基于用户的完美调度多路访问

user destination routing (UDR)　用户目标路由选择

user element (UE)　用户元素

user equipment (UE)　用户设备

user experience (UE)　用户体验

user experience design competency　用户体验设计能力

user experience design manager　用户体验设计经理

user experience design project　用户体验专题设计

user experience design resources　用户体验设计资源

user facing-provider edge (UPE)　面向用户运营商边缘设备

user generated content (UGC)　用户生成内容

user identification code (UIC)　用户标识码

user identity model (UIM)　用户识别模块

user interface program (UIP)　用户接口程序

user level protocol　用户级协议

user-level security　用户级安全性

user log off　用户注销

user log on　用户注册

user name　用户名

user network behavior analysis　用户网络行为分析

user network interface (UNI)　用户网络接口

user network interface management entity (UME)　用户网络接口管理实体

user node　用户节点

user port　用户端口

user port function (UPF)　用户端口功能

user protocol conversion　用户协议转换

user to user signaling (UUS)　用户-用户信令

user-user protocol　用户-用户协议

USF (uplink state flag)　上行链路状态标记

USN (ubiquitous sensor network)　泛在传感器网络

USSD (unstructured supplementary service data)　非结构化的补充数据业务

usual asymmetric digital subscriber line (UADSL)　通用非对称数字用户线路

UTM (united threat management)　统一威胁管理

UTP (unshielded twisted pair)　非屏蔽双绞线

UTRAN (UMTS terrestrial radio access network)　UMTS陆地无线接入网，通用移动通信系统陆地无线接入网

UUID (universal unique identifier)　通用唯一标识符

UUS (user to user signaling)　用户-用户信令

UV (unique visitor)　独立访客

UWB (ultra wideband)　超宽带

# V

VACC (value added common carrier) 增值公用载波
value-added common carrier (VACC) 增值公用载波（公司）
value-added network (VAN) 增值网
value-added network service (VANS) 增值网络业务
value-added service (VAS) 增值业务
VAN (virtual application network) 虚拟应用网络
VAN (value-added network) 增值网
VAN gateway 增值网网关
VANS (value-added network service) 增值网络业务
variable bit rate (VBR) 可变位速率
variable length subnet mask (VLSM) 可变长度子网掩码
variable rate adaptive multiplexing 可变速率自适应多路复用
VAS (value-added service) 增值业务
VBAS (virtual broadband access server) 虚拟宽带接入服务器
VBR (variable bit rate) 可变位速率
VBS (voice broadcast service) 话音广播业务
VC (virtual circuit) 虚拟电路
VC (virtual channel) 虚拟通道
VC (virtual concatenation) 虚级联

VC (virtual container) 虚容器
VCC (virtual call capability) 虚呼叫能力
VCC (virtual channel connection) 虚拟通道连接
VCCC (virtual circuit congestion control) 虚拟电路拥塞控制
VCG (virtual concatenation group) 虚级联组
VCH (virtual channel handler) 虚信道处理器
VCI (virtual chain index) 虚链路索引
VCI (virtual channel identifier) 虚拟通道标识符
VCL (virtual channel link) 虚拟通道链路
VCL (virtual CD-ROM library) 虚拟光碟库
VCN (virtual call network) 虚呼叫网络
VCS (virtual content server) 虚拟内容服务器
V-D algorithm 向量距离算法
VDHA (virtual dynamic hierarchical architecture) 虚拟动态分层体系结构
VDSL (very high-bit-rate digital subscriber line) 超高位速率数字用

户线
vector distance 矢量距离
vector distance routing 矢量距离路由选择
vehicle Ad hoc network 车载自组织网络
velocity of propagation (VOP) 传播速度
VE-mail (voice E-mail) 语音电子邮件
versatile message transaction protocol (VMTP) 通用报文事务协议
vertical portal 垂直门户
vertical search engine (VSE) 垂直搜索引擎
vertical wiring subsystem 垂直布线子系统
very high-bit-rate digital subscriber line (VDSL) 超高位速率数字用户线路
very severe burst (VSB) 极严重突发脉冲串
vestigial sideband (VSB) 残留边带
vestigial sideband modulation (VSM) 残留边带调制
vestigial sideband transmission 残留边带传输
VG (very good) 很好
VGC (videotex gateway control) 可视图文网关控制
VGMP (VRRP group management protocol) VRRP 组管理协议
VHE (virtual home environment) 虚拟家乡环境
video communication network 视频通信网络
video conferencing service (VCS) 视频会议业务
video conferencing system (VCS) 视频会议系统
video mail system (VMS) 视频邮件系统
video on demand (VOD) 视频点播
video sharing (VS) 视频共享
videotex gateway control (VGC) 可视图文网关控制
video transcoding 视频转码
video transmission service (VTS) 视频传输业务
VIEID (virtual identity electronic identification) 虚拟身份电子标识
VIEW (virtual interaction environment workstation) 虚拟交互环境工作站
V interface V 接口
viral marketing 病毒营销
virtual access method 虚拟存取法
virtual active network 虚拟主动网络
virtual address (VA) 虚拟地址
virtual analogue switching point (VASP) 虚拟模拟交换点
virtual application network (VAN) 虚拟应用网络
virtual dialing private network 虚拟拨号专用网
virtual dynamic hierarchical architecture (VDHA) 虚拟动态分层体系结构
virtual bridged local area networks 虚拟桥接局域网
virtual broadband access server (VBAS) 虚拟宽带接入服务器
virtual call 虚呼叫
virtual call capability (VCC) 虚呼叫能

力
virtual call facility 虚拟呼叫设施
virtual call mode 虚呼叫模式
virtual call network（VCN） 虚呼叫网络
virtual call service 虚呼叫服务
virtual community 虚拟社区
virtual concatenation（VC） 虚级联
virtual concatenation group（VCG） 虚级联组
virtual CD-ROM library（VCL） 虚拟光碟库
virtual chain index（VCI） 虚链路索引
virtual channel（VC） 虚拟通道
virtual channel connection（VCC） 虚拟通道连接
virtual channel handler（VCH） 虚拟通道处理器
virtual channel identifier（VCI） 虚拟通道标识符
virtual channel level 虚信道级
virtual channel link（VCL） 虚拟通道链路
virtual channel switch 虚拟通道交换器
virtual circuit（VC） 虚拟电路
virtual circuit congestion control（VCCC） 虚拟电路拥塞控制
virtual circuit connection 虚拟电路连接
virtual circuit descriptor 虚拟电路描述符
virtual circuit number 虚拟电路编号
virtual circuit pacing 虚拟电路调步
virtual cloud 虚云,虚拟云

virtual community 虚拟社区
virtual computing 虚拟计算
virtual computing environment（VCE） 虚拟计算环境
virtual concatenation 虚拼接
virtual connection 虚连接
virtual container（VC） 虚容器
virtual content server（VCS） 虚拟内容服务器
virtual currency 虚拟货币
virtual currency capital 虚拟货币资本
virtual cut through 虚拟虚探路
virtual datagram 虚数据报
virtual data network 虚拟数据网
virtual digital local area network 虚拟数字局域网
virtual directory 虚拟目录
virtual economy 虚拟化经济
virtual Ethernet port aggregator 虚拟以太网端口聚合器
virtual extensible local area network 虚拟可扩展局域网
virtual file server 虚拟文件服务器
virtual file store 虚拟文件存储器
virtual home environment（VHE） 虚拟家乡环境
virtual host 虚拟主机,虚拟服务器
virtual identity electronic identification（VIEID） 虚拟身份电子标识,电子标识
virtual interaction environment workstation（VIEW） 虚拟交互环境工作站
virtual leased line（VLL） 虚拟租用线
virtual link packet switching 虚拟链路包交换

virtual local area network (VLAN) 虚拟局域网
virtual local area network hopping attack 虚拟局域网跳跃攻击
virtual local area network multicast 虚拟局域网组播
virtual local area network technology 虚拟局域网技术
virtual machine identifier 虚拟机标识符
virtual machine interface 虚拟机接口
virtual market 虚拟市场
virtual media gateway (VMG) 虚拟媒体网关
virtual MIMO 虚拟多入多出
virtual MIMO system 虚拟多输入多输出系统
virtual MIMO wireless sensor network (VMWSN) 虚拟多天线无线传感器网络
virtual network 虚拟网
virtual network operator (VNO) 虚拟网络运营商
virtual network switching 虚拟网络切换
virtual organization 虚拟组织
virtual path (VP) 虚拟通道
virtual path connection (VPC) 虚拟路径连接
virtual path connection identifier 虚拟路径连接标志符
virtual path identifier (VPI) 虚拟路径标识符
virtual path level 虚通路级
virtual path link (VPL) 虚拟路径链路
virtual path switch 虚拟路径交换器
virtual path terminator (VPT) 虚拟路径终端器
virtual private cloud (VPC) 虚拟私有云
virtual private data network 虚拟私用数据网
virtual private dial-up network (VPDN) 虚拟专用拨号网
virtual private LAN service (VPLS) 虚拟专用局域网业务
virtual private line 虚拟专用线
virtual private mobile network (VPMN) 虚拟专用移动网络
virtual private network (VPN) 虚拟专用网
Virtual Private Network Consortium (VPNC) 虚拟专用网协会
virtual private network forward gateway 虚拟专用网转发网关
virtual private network service (VPNS) 虚拟专用网业务
virtual private remote network 虚拟专用远程网
virtual private routed network (VPRN) 虚拟专用路由网
virtual private wire service (VPWS) 虚拟专用线路业务
virtual reality (VR) 虚拟现实
virtual reality gaming 虚拟现实游戏
virtual reality online gaming 虚拟现实网络游戏
virtual reality technology 虚拟技术
virtual reality world 虚拟世界
virtual resource research 虚拟资源

研究

virtual route 虚拟路由

virtual route identifier (VRID) 虚拟路由标识符

virtual route pacing 虚拟路由调步

virtual route pacing request (VRPRQ) 虚拟路由调步请求

virtual route pacing response (VRPRS) 虚拟路由调步响应

virtual router (VR) 虚拟路由器

virtual router redundancy protocol (VRRP) 虚拟路由器冗余协议

virtual route sequence number 虚拟路由顺序号

virtual routing node 虚拟路由节点

virtual scaleable network 虚拟可伸缩网络

virtual scheduling (VS) 虚拟调度

virtual server 虚拟服务器

virtual server architecture (VSA) 虚拟服务器架构

virtual shop 虚拟商店,网上商店

virtual society 虚拟社会

virtual source (VS) 虚拟源端

virtual storefront 虚拟商店

virtual telecommunication access method (VTAM) 虚拟远程通信访问法

virtual terminal (VT) 虚拟终端

virtual terminal environment 虚拟终端环境

virtual terminal network 虚拟终端网络

virtual terminal protocol (VTP) 虚拟终端协议

virtual terminal voice band line (VTVBL) 虚拟终端音频线路

virtual tributary (VT) 虚支路

virtual university 虚拟大学

virtual video sharing network platform (VVSNP) 虚拟视频共享网络平台

virtual wavelength path (VWP) 虚波长通道

virtual Wi-Fi router 虚拟无线路由器

visited location register (VLR) 漫游位置寄存器

visited mobile service switching center (VMSC) 受访移动交换中心

VLAN (virtual local area network) 虚拟局域网

VLAN hopping attack 虚拟局域网跳跃攻击

VLL (virtual leased line) 虚拟租用线

VLR (visited location register) 漫游位置寄存器

VLSM (variable length subnet mask) 可变长度子网掩码

VMG (virtual media gateway) 虚拟媒体网关

VMS (video mail system) 视频邮件系统

VMS (voice mail system) 语音邮件系统

VMSC (visited mobile service switching center) 受访移动交换中心

VMTP (versatile message transaction protocol) 通用报文事务协议

VMWSN (virtual MIMO wireless sensor network) 虚拟多天线无线传感器网络

VNO (virtual network operator) 虚拟

网络运营商
voice broadcast service (VBS)　话音广播业务
voice control service (VCS)　语音控制服务
voice E-mail (VE-mail)　语音电子邮件
voice gateway (VG)　语音网关
voice mail　语音邮件
voice mail system (VMS)　语音邮件系统
voice navigation　语音导航
voice on the net (VON)　网上语音
voice over Internet protocol (VoIP)　网络电话，IP电话
voice store and forward　语音存储转发
VoIP (voice over Internet protocol)　网络电话，IP电话
VON (voice on the net)　网上语音
VP (virtual path)　虚拟通道
VPC (virtual path connection)　虚拟路径连接
VPC (virtual private cloud)　虚拟私有云
VP cross connect　虚拟通道交叉连接
VPDN (virtual private dial-up network)　虚拟专用拨号网
VPI (virtual path identifier)　虚拟路径标识符
VPL (virtual path link)　虚拟路径链路
VPLS (virtual private LAN service)　虚拟专用局域网业务
VPMN (virtual private mobile network)　虚拟专用移动网络
VPN (virtual private network)　虚拟专用网

VPNC (Virtual Private Network Consortium)　虚拟专用网协会
VPRN (virtual private routed network)　虚拟专用路由网
VP switch　虚拟通道交换
VPT (virtual path terminator)　虚拟路径终端器
VPWS (virtual private wire service)　虚拟专用线路业务
VR (virtual reality)　虚拟现实
VR (virtual route)　虚拟路由
VRID (virtual route identifier)　虚拟路由标识符
VRPRQ (virtual route pacing request)　虚拟路由调步请求
VRPRS (virtual route pacing response)　虚拟路由调步响应
VRRP (virtual router redundancy protocol)　虚拟路由器冗余协议
VRRP group management protocol (VGMP)　虚拟路由器冗余协议组管理协议
VS (virtual source)　虚拟源端
VSA (virtual server architecture)　虚拟服务器架构
VSB (very severe burst)　极严重突发脉冲串
VSB (vestigial sideband)　残留边带
VSE (vertical search engine)　垂直搜索引擎
VSM (vestigial sideband modulation)　残留边带调制
VT (virtual terminal)　虚拟终端
VT (virtual tributary)　虚支路
VTAM (virtual telecommunication access

method) 虚拟远程通信访问法
VTOA (voice and telephone over ATM) 基于 ATM 的语音和电话传输
VTP (virtual terminal protocol) 虚拟终端协议

VVSNP (virtual video sharing network platform) 虚拟视频共享网络平台
VWP (virtual wavelength path) 虚波长通道

# W

WAD (Web Applications Development) 网络应用程序开发

WADS (wide area data service) 大范围数据服务

WAE (wireless application environment) 无线应用环境

WAF (Web application firewall) Web 应用防火墙

WAIS (wide area information service) 广域信息服务

wake on local area network 局域网唤醒

walled garden 围墙花园

WAN (wide area network) 广域网

WAN call destination 广域网呼叫目标

WANET (wireless Ad hoc network) 无线自组织网络

Wang net 王安网

WAP (wireless application protocol) 无线应用协议

WAP (wireless access point) 无线接入点

WAP (Web application proxy) Web 应用程序代理

WAS (Web application server) Web 应用服务器

WAU (weekly activited users) 周活跃用户

wave division multiplexing (WDM) 波分复用

wave division multiplexing passive optical network (WDM-PON) 波分复用无源光网络

wavelength conversion 波长转换

wavelength converter (WC) 波长转换器

wavelength cross connect (WXC) 波长交叉连接

wavelength division multiple access (WDMA) 波分多址

wavelength interchanging cross connector (WIXC) 波长变换交叉连接器

wavelength path (WP) 波长通道

wavelength router 波长路由器

wavelength routing optical network (WRON) 波长路由光网络

wavelength routing switching (WRS) 波长路由交换

wavelength selective cross connector (WSXC) 波长选择交叉连接器

wavelet fuzzy neural network 小波模糊神经网络

WBAN (wireless body area network) 无线体域网

WBC (wide band channel) 宽带信道

WBDL（wide band data link） 宽带数据链路
WBEM（Web-based enterprise management） 基于 Web 的企业管理标准
WBM（Web based management） 基于 Web 的网络管理
WC（wavelength converter） 波长转换器
WCDMA（wideband code division multiple access） 宽带码分多址
WCS（Web coverage service） Web 覆盖服务,网络覆盖服务
WDA（wireless digital assistant） 无线数字助理
WDM（wave division multiplexing） 波分复用
WDM all optical communication network 波分复用全光通信网
WDMA（wavelength division multiple access） 波分多址
WDM-PON（wave division multiplexing passive optical network） 波分复用无源光网络
WDP（wireless datagram protocol） 无线数据报协议
WDS（wireless distribution system） 无线分布式系统
WDTP（wireless dynamic token protocol） 无线动态令牌协议
weakly guiding fiber（WGF） 弱导光纤
Web 网,网页,网站,万维网
Web address Web 地址
Web analysis 网页分析
Web anonymizer 网络匿名
Web Applications Development（WAD） 网络应用程序开发
Web application firewall（WAF） 网络应用防火墙
Web application proxy（WAP） Web 应用程序代理
Web application server（WAS） Web 应用服务器
Web asynchronous communication Web 异步通信
Web-based application 基于 Web 应用
Web-based enterprise management（WBEM） 基于 Web 的企业管理标准
Web-based management（WBM） 基于 Web 的网络管理
Web-based project management 基于 Web 的项目管理
Web broker Web 代理
Web browser Web 浏览器
Web caching 网络缓存,Web 缓存
Web caching server Web 缓存服务器
Webcast 网络直播,网播
Webcast services 网络直播服务
Webciety 网络社会
Web conference 网上会议
Web conferencing service 网络会议服务
Web content distributor 万维网内容分发商
Web content filtering Web 内容过滤
Web content mining 万维网内容挖掘
Web coverage service（WCS） Web 覆盖服务
Web crawler 网络爬虫
Web crowdfunding Web 众筹

Web database application　Web 数据库应用
Web debugging proxy　Web 调试代理
web defacement　网页涂改
Web developing application　Web 应用开发
Web economics　Web 经济
Web farm　Web 场,Web 服务器场,网络场
Web farming　Web 耕作
Web filtering proxy server(WFPS)　网页过滤代理服务器
Web friendly link　网站友情链接
Web fuzzy clustering　Web 模糊聚类
Web game　网页游戏
Web geographic information system　Web 地理信息系统
Web harvesting　Web 收割
Web hosting　网站托管
web hosting infrastructure　网站托管基础设施
web hosting model　网站托管模型
web hosting provider　网站托管提供商
Web image annotation　Web 图像标注
Web index　Web 索引
Web information filtering　Web 信息过滤
Web link　Web 链接
Web link analysis　Web 链接分析
Web link exchange　Web 链接互换
Web links filtering　Web 链接过滤
Web link URL　Web 链接地址
Weblogger　博客
Web mail service　Web 邮件服务
Web map service　Web 地图服务

Web media　Web 媒体
Webmetrics　Web 信息计量学
Web mining　网站挖掘
Web mining application　网站数据挖掘应用
Webnomics　Web 经济
Web offline application　Web 离线应用
Web ontology language　Web 本体语言
Web oriented architecture(WOA)　面向 Web 的架构
Web page　网页
Web page acceleration　网页加速
Web page categorization　网页分类
Web page code　网页代码,Web 页面代码
Web page firewall　网页防火墙
Web page link　网页链接
Web page malicious code　网页恶意代码
Web page rank　网页排名
Web page search engine　网页搜索引擎
Web personalization　网站个性化
Web phone　Web 电话
Web portal　门户网站,网络门户
Web portal manager　网络门户管理(软件)
Web property　Web 资产
Web proxy　Web 代理
Web proxy auto-discovery　Web 代理自动发现
Web public opinion　网络舆情
Web quest　网络探究
Web real time communication(WebRTC)　Web 实时通信
Web request broker(WRB)　Web 请求

代理

Web reverse proxy　Web 反向代理

Web risk　Web 风险

WebRTC（Web real time communication）　Web 实时通信

Web search engine　Web 搜索引擎

Web search service　Web 搜索服务

Web security technology　Web 安全技术

Web self-service（WSS）　Web 自助式服务

Web semantic cluster　Web 语义簇

Web server　Web 服务器，网站服务器

Web server application program　Web 服务器应用程序

Web server caching　Web 服务器缓存

Web server clustering　Web 服务器集群

Web server control　Web 服务器控件

Web server farm　Web 服务器场

Web service（WS）　Web 服务，网络服务

Web service architecture　Web 服务架构

Web service-based　基于 Web 服务

Web service-based grid　基于 Web 服务网格

Web service-based integration　基于 Web 服务的集成

Web service-based management services　基于 Web 服务的管理服务

Web service capabilities　Web 服务能力

Web service description language（WSDL）　Web 服务描述语言

Web service flow language（WSFL）　Web 服务流语言

Web service for remote portal（WSRP）　远程门户 Web 网络服务

Web service in cloud　云内 Web 服务

Web service matching　Web 服务匹配

Web service resource framework　Web 服务资源框架

website　网站

website code　网站代码

website construction　网站建设

website design　网站设计

website development　网站开发

website editor　网站编辑

website embedded Trojan　网站嵌入特洛伊木马

website hang a horse　网站挂马

website hits　点击数

website homepage　网站首页

website maintenance　网站维护

website navigation　网站导航

website of resource information　资源信息网站

website promotion　网站推广

website security　网站安全

website traffic analysis　网站流量分析

website traffic rank　网站流量排名

website traffic statistic　网站流量统计

Web spider　网络蜘蛛

Web storefront　网上商店，网店

Web terminal　Web 终端

Web text filtering　Web 文本过滤

Web title　网页标题

Web tunnel　Web 隧道技术

Web TV　Web 电视

Web user authentication server　Web 用户身份验证服务器

Web user login server　Web 用户登录服务器
Web user server　Web 用户服务器
Web writer　Web 写手
Webzine (Web magazine)　Web 杂志
WeChat　微信
WeChat advertising　微信广告
WeChat crowdfunding　微信众筹
WeChat friend circle　微信朋友圈
WeChat follow　微信关注
WeChat ID　微信号
WeChat marketing　微信营销
WeChat moments　微信朋友圈
WeChat official accounts　微信公众号
WeChat pay　微信支付
WeChat promotion　微信推广
WeChat public accounts　微信公众号
WeChat public number　微信公众号
WeChat public platform　微信公众平台
Wechat pyramid selling　微信传销
Wechat red packet　微信红包
Wechat small store　微商
WeChat subions　微信订阅号
WeChat wallet　微信钱包
WeChat Webpage two-dimensional code　微信网页二维码
weekly activited users (WAU)　周活跃用户
weighted fair queuing (WFQ)　加权公平排队
well-known host name　知名宿主名
well-known ports　知名端口，公认端口
We Media　自媒体
We Media entertainment　自媒体娱乐

We media era　自媒体时代
WFMC (workflow management coalition)　工作流管理联盟
WFQ (weighted fair queuing)　加权公平排队
WFPS (Web filtering proxy server)　网页过滤代理服务器
WGF (weakly guiding fiber)　弱导光纤
what you see is what you get　所见即所得
white board　白板
white board service　白板服务
white hat Hacker　白帽黑客
white hat SEO　白帽 SEO, 白帽搜索引擎优化
white page　白页
who-are-you (WRU)　你是谁, WRU 信号
Whois client　Whois 客户机程序
Whois server　Whois 服务器程序
whole site accelerator (WSA)　全站加速
wide area communication　广域通信
wide area communication system　广域通信系统
wide area computing service　广域计算服务
wide area data service (WADS)　大范围数据服务
wide area gigabit network　广域千兆位网络
wide area information network hub　广域信息网络集线器
wide area information service (WAIS)　广域信息服务

wide area information transfer system 广域信息传送系统

wide area message system 广域报文系统，广域消息系统

wide area network (WAN) 广域网

wide area network interface module 广域网接口模块

wide area remote sensor 广域远程传感器

wide area subnet 广域子网

wide area switch fabric 广域交换结构

wide area switch fiber network 广域交换光纤网

wide area telecommunication service 广域通信业务

wide area telephone service (WATS) 广域电话业务

wide area transmission system 广域传输系统

wide area virtual computing 广域虚拟计算

wide band channel (WBC) 宽带信道

wideband code division multiple access (WCDMA) 宽带码分多址

wide band data link (WBDL) 宽带数据链路

wide band packet technology 宽带分组技术

wide band ratio 宽带比

wide network interface module (WNIM) 广域网接口模块

wide sense non-blocking network 广义无阻塞网络

Widget 互联精灵

Wi-Fi (wireless fidelity) 无线保真，无线网络

Wi-Fi hotspot 无线热点

Wi-Fi router 无线路由器

Wi-Fi protected access (WPA) Wi-Fi网络安全存取

wiki 维客

wikipedia 维基百科

WiMAX (worldwide interoperablity for microwave access) 微波接入全球互通

WIN (wireless intelligent network) 无线智能网

Windows distributed internet applications architecture (Windows DNA) Windows 分布式集成网络应用体系结构

Windows DNA (Windows distributed internet applications architecture) Windows 分布式集成网络应用体系结构

Windows Internet naming service (WINS) Windows 因特网命名服务

Windows NT WindowsNT 操作系统

Windows NT advanced server WindowsNT 高级服务器版

Windows open service architecture (WOSA) Windows 开放式服务体系结构

Windows open system architecture (WOSA) Windows 开放系统体系结构

Windows sockets (Winsock) Windows 系统套接字接口

WINS (Windows Internet naming service) Windows 因特网命名服务

wired equivalent privacy (WEP) 有线

等效加密

wireless access　无线接入

wireless access network　无线接入网

wireless access point（WAP）　无线接入点

wireless access service　无线接入业务

wireless Ad hoc network（WANET）　无线自组织网络

wireless application environment（WAE）　无线应用环境

wireless application protocol（WAP）　无线应用协议

wireless asynchronous Web services　无线异步 Web 服务

wireless ATM（WATM）　无线 ATM，无线异步传输模式

wireless band selective repeater　无线选带直放站

wireless body area network（WBAN）　无线体域网

wireless customer premises network　无线用户驻地网络

wireless datagram protocol（WDP）　无线数据报协议

wireless digital assistant（WDA）　无线数字助理

wireless digital local loop　无线数字本地环路

wireless distribution system（WDS）　无线分布式系统

wireless dynamic token protocol（WDTP）　无线动态令牌协议

wireless fidelity（Wi-Fi）　无线保真

wireless hub　无线网络集线器

wireless integrated multiple access　无线综合多址接入

wireless intelligent network（WIN）　无线智能网

wireless Internet of things　无线物联网

wireless IP sub layer　无线 IP 子层

wireless LAN（wireless local area network）　无线局域网

wireless LAN switch　无线局域网交换机

wireless local area network（WLAN）　无线局域网

wireless local loop（WLL）　无线本地环路

wireless local loop unit　无线本地环路单元

wireless markup language（WML）　无线标记语言

wireless mesh backbone network　无线网状骨干网络

wireless mesh backhaul network　无线网状数据回程网

wireless mesh network（WMN）　无线网状网络

wireless mesh router　无线网状路由器

wireless mesh technology　无线网状网技术

wireless metropolitan area network（WMAN）　无线城域网

wireless mobile Ad hoc networks　无线移动 Adhoc 网络

wireless mobile communication networks　无线移动通信网络

wireless mobile self-organization network　无线移动自组织网络

wireless mobile sensor networks　无线

移动传感器网络
wireless mobile network 无线移动网络
wireless multihop mobile network 无线多跳移动网络
wireless network design 无线网络设计
wireless network equipment (WNG) 无线网络设备
wireless network guards (WNG) 无线网络卫士
wireless network interface 无线网络接口
wireless network location 无线网络定位
wireless network planning 无线网络规划
wireless network security 无线网络安全
wireless office system (WOS) 无线办公系统
wireless personal area network (WPAN) 无线个人区域网
wireless personal communication (WPC) 无线个人通信
wireless public key infrastrcture (WPKI) 无线公开密钥体系
wireless regional area network (WRAN) 无线区域网络
wireless repeater 无线中继
wireless roaming 无线漫游
wireless router 无线路由器
wireless router array 无线路由器阵列
wireless sensor network (WSN) 无线传感器网络
wireless sensor network coverage 无线传感器网络覆盖
wireless session protocol (WSP) 无线会话协议
wireless sensor surveillance network (WSN) 无线传感器监测网络
wireless sensor tracking network (WSN) 无线传感器追踪网络
wireless telephony application (WTA) 无线电话应用
wireless terminal 无线终端
wireless transaction control protocol (WTCP) 无线传输控制协议
wireless transaction protocol (WTP) 无线传输协议
wireless transaction protocol specification 无线传输协议规范
wireless transport layer security (WTLS) 无线传输层安全
wireless USB (WUSB) 无线 USB
wireless virtual LAN 无线虚拟局域网
wireless virtual private network (WVPN) 无线虚拟专用网
wireless wide area network (WWAN) 无线广域网
wisdom of the earth 地球的智慧
witkey 威客
witkey marketing 威客营销
witkey platform 威客平台
Witkey website 威客网站
WIXC (wavelength interchanging cross connector) 波长变换交叉连接器
WLAN (wireless local area network) 无线局域网
WLAN (wake on local area network) 网络唤醒

WLL (wireless local loop) 无线本地环路
WMAN (wireless metropolitan area network) 无线城域网
WML (wireless markup language) 无线标记语言
WMN (wireless mesh network) 无线网状网络
WNG (wireless network guards) 无线网络卫士
WNIM (wide network interface module) 广域网接口模块
WOA (Web oriented architecture) 面向Web的架构
word-of-mouth communication 口碑传播
workflow 工作流
workflow designer 工作流设计器
workflow engine 工作流引擎
workflow management coalition (WFMC) 工作流管理联盟
workflow management system (WFMS) 工作流管理系统
workgroup 工作组
worldwide interoperablity for microwave access (WiMAX) 微波接入全球互通
worldwide network agent 全球网络代理
world wide web (WWW) 万维网
World Wide Web Consortium (W3C) 万维网联盟
World Wireless Research Forum (WWRF) 世界无线研究论坛
wormhole 虫孔
WOS (wireless office system) 无线办公系统
WOSA (Windows open service architecture) Windows开放式服务体系结构
WOSA (Windows open system architecture) Windows开放系统体系结构
WP (wavelength path) 波长通道
WPA (Wi-Fi protected access) Wi-Fi网络安全存取
WPAN (wireless personal area network) 无线个人区域网
WPC (wireless personal communication) 无线个人通信
WPKI (wireless public key infrastrcture) 无线公开密钥体系
WRAN (wireless regional area network) 无线区域网络
WRB (Web request broker) Web请求代理
WRON (wavelength routing optical network) 波长路由光网络
WRS (wavelength routing switching) 波长路由交换
WS (Web service) Web服务
WSA (whole site accelerator) 全站加速
WSDL (Web service description language) Web服务描述语言
WSFL (Web service flow language) Web服务流语言
WSN (wireless sensor network) 无线传感器网络
WSP (wireless session protocol) 无线会话协议
WSRP (Web service for remote portal)

远程门户 Web 网络服务

WSS (Web self-service) 网络自助式服务

WSXC (wavelength selective cross connector) 波长选择交叉连接器

WTA (wireless telephony application) 无线电话应用

WTCP (wireless transaction control protocol) 无线传输控制协议

WTLS (wireless transport layer security) 无线传输层安全

WTP (wireless transaction protocol) 无线传输协议

WUSB (wireless USB) 无线 USB

WVPN (wireless virtual private network) 无线虚拟专用网

WWAN (wireless wide area network) 无线广域网

WWRF (World Wireless Research Forum) 世界无线研究论坛

WWW (World Wide Web) 万维网

WWW dynamical access 万维网动态访问

WWW static access 万维网静态访问

WXC (wavelength cross connect) 波长交叉连接

W3 (World Wide Web) W3 网

W3C (World Wide Web Consortium) 万维网联盟

# X

XAUI (10 gigabit attachment unit interface) 10兆位附件单元接口,XAUI接口

Xerox network service (XNS) 施乐网络服务

Xerox network system (XNS) 施乐网络系统

Xerox network system/internet transport 施乐网络系统/网间传送

Xerox network systems' Internet transport protocol (XNS/ITP) 施乐网络系统的互联网传输协议

X interface X接口

XLink (XML linking language) XML文本链接

XLL (extensible link language) 可扩展链接语言

XMI (XML metadata interchange) XML元数据交换

XML (extensible markup language) 可扩展标记语言

XML metadata interchange (XMI) XML元数据交换

XML section XML节

XML stream XML流

Xmodem Xmodem协议

Xmodem-CRC 带CRC校验的Xmodem协议

XMPP (extensible messaging and presence protocol) 可扩展消息和呈现协议

XNS (Xerox network system) 施乐网络系统

XNS (Xerox network service) 施乐网络服务

XNS network layer protocols XNS网络层协议

X series recommendations X系列建议

XSL (extensible stylesheet language) 可扩展样式语言

XSS (cross site scripting) 跨站脚本

XTACACS (extended terminal access contorller access control system) 扩展终端访问控制器访问控制系统

XTP (express transfer protocol) 快速传输协议

X.25 interface X.25接口

X.25 NCP packet switching interface (NPSI) X.25网络控制程序分组交换接口

X.25 packet switch data network X.25分组交换数据网

X.25 protocol X.25协议

X.25 recommendation X.25建议

# Y

**YB（yottabyte）** 尧字节
**yellow book** 黄皮书
**yellow pages** 黄页
**YHBT（You have been trolled）** 你被钓到了
**YHL（You have lost）** 你上当了
**Ymodem protocol** Ymodem 协议
**Ymodem-g** Ymodem-g 协议
**yotta** 尧它
**yottabyte（YB）** 尧字节

# Z

ZB (zettabyte) 泽字节
ZDA (zero-day attack) 零时差攻击
ZDL (zero delay lockout) 零延迟锁定
zero bit insertion 零位插入
zero bit time-slot insertion (ZBTSI) 零位时隙插入
zero committed information rate (ZCIR) 零承诺信息速率
zero-day attack (ZDA) 零时差攻击
zero delay lockout (ZDL) 零延迟锁定
zero insertion 零插入
zero modulation (ZM) 调零,零调制
zero-slot LAN 零插槽局域网
zero transfer function 零传输函数
zetta 泽它
zettabyte (ZB) 泽字节
ZigBee ZigBee 技术,紫蜂技术
ZigBee alliance ZigBee 联盟
ZIP (zone information protocol) 区域信息协议
ZIS (zone information socket) 区域信息套接口
ZIT (zone information table) 区域信息表
ZM (zero modulation) 调零,零调制
Zmodem Zmodem 协议
Z-net Z-网
zone information protocol (ZIP) 区域信息协议
zone information socket (ZIS) 区域信息套接口
zone information table (ZIT) 区域信息表

# 以数字起首的词条

1 base-5　1兆位基带5网
1-persistent CSMA　1率持续载波监听多路访问
10 base-2　10兆位基带2网
10 base-5　10兆位基带5网
10 base-F　10兆位基带F网
10 base-FB　10兆位基带FB网
10 base-FL　10兆位基带FL网
10 base-FP　10兆位基带FP网
10 base-T　10兆位基带T网
10 broad-36　10兆位宽带36网
10 gigabit attachment unit interface (XAUI)　10兆位附件单元接口，XAUI接口
10 gigabit media independent interface (XGMII)　10兆位独立于介质的接口，XGMII接口
100 base-FX　100兆位基带FX网
100 base-T　100兆位基带T网
100 base-T technical features　100兆位基带T网技术特点
100 base-TX　100兆位基带TX网
100 base-T4　100兆位基带T4网
100 base-VG　100兆位基带VG网
100 base-X　100兆位基带X网
100 VG-AnyLAN　100兆位语音级网
1 000 base-CX　1000兆位基带CX网
1 000 base-LX　1000兆位基带LX网
1 000 base-SX　1000兆位基带SX网
1 000 base-T　1000兆位基带T网
10 000 base　万兆位基带网
2B1Q　2B1Q码
3A (Authentication, Authorization, Accounting)　认证，授权，计费
3A revolution　3A革命
3C (Computer, Communication, Consumer)　计算机＋通信＋消费者
3D virtual reality technique　三维虚拟现实技术
3G (third generation mobile system)　第三代(移动通信系统)
3GPP (third generation partnership project)　第三代移动通信项目组织
3GPP2 (third generation partnership project 2)　第三代移动通信项目组织2
3G multimedia streaming services　3G多媒体流业务
3-network fusion　三网融合
3 tiered network architecture　三层网络架构
3W (World Wide Web)　万维网
301 redirect　301重定向，永久重定向
302 redirect　302重定向，暂时重定向
4A (Authentication, Account, Authorization, Audit)　认证，账号，授权，审计

**4A communication**　4A 通信
**4B/5B coding**　4B/5B 编码
**4B/5B local fiber**　4B/5B 局部光纤
**4G (fourth generation mobile communication technology)**　第四代移动通信技术
**5-4-3 rule**　5-4-3 规则
**568A standard**　双绞线布线标准 568A
**568B standard**　双绞线布线标准 568B
**6 to 4 address**　6to4 地址
**6 to 4 tunneling**　6to4 隧道

**8B/10B coding**　8B/10B 编码
**8B/10B local fiber**　8B/10B 局部光纤
**8 phase shift keying (8PSK)**　8 移相键控
**8PSK (8 phase shift keying)**　8 移相键控
**802.3 frame structure**　802.3 帧格式
**802.3 MAC sublayer protocol**　802.3 介质访问控制子层协议